JOURNEY
TO THE
STARS

ROBERT JASTROW

JOURNEY TO THE STARS

Space Exploration–
Tomorrow and Beyond

BANTAM BOOKS
NEW YORK · TORONTO · LONDON · SYDNEY · AUCKLAND

JOURNEY TO THE STARS
A Bantam Book / September 1989

Library of Congress Cataloging-in-Publication Data

Jastrow, Robert, 1925–
    Journey to the stars : space exploration, tomorrow and beyond /
Robert Jastrow.
        p.   cm.
    Includes index.
    ISBN 0-553-05386-8 ISBN 978-0-553-34909-2 (pbk.)
    1. Outer space—Exploration. 2. Solar system. 3. Stars.
4. Astronautics in astronomy. I. Title.
QB500.J37   1989
919.904—dc20   89-31920
                                                                         CIP

Bantam Books are published by Bantam Books, a division of Bantam
Doubleday Dell Publishing Group, Inc. Its trademark, consisting of the
words "Bantam Books" and the portrayal of a rooster, is Registered in
U.S. Patent and Trademark Office and in other countries. Marca
Registrada. Bantam Books, 666 Fifth Avenue, New York, New York
10103.

BVG 01
146911954

To my mother
and to the memory of
my father

Contents

four and a half billion years old. It follows that if other solar systems contain intelligent life, this life must, on the average, have evolved at least a billion years beyond mankind.

Reflecting on the fact that the human brain has doubled in size in the last million years, the astronomers say to themselves: Beings whose evolution has carried them a billion years beyond the human level may possess great heights of brainpower; how interesting it would be to speak to them, and find out what they have learned in that long interval. The greatest thinkers in human history might be as children in their presence. "In their eyes," one observer suggests, "Einstein would qualify as a waiter and Thomas Jefferson as a busboy."

How long must we wait before we are admitted to the company of these stimulating creatures? If the extraterrestrials exist, our television signals have made them aware of our presence; but perhaps they find our thoughts too simple to warrant a response. Perhaps another billion years must pass before the mature beings of the Cosmos judge mankind's descendants worthy of their attention.

It seems a pity that we must wait so long for the interesting exchange to begin. But this reasoning may underestimate the curiosity the advanced extraterrestrials might display in such new arrivals as Homo sapiens. Intelligence and curiosity go together; the discovery of new worlds may be life's greatest pleasure for a highly intelligent being. It is difficult to avoid a suspicion that the advanced beings—if they exist—have visited the solar system at least once in all those billions of years. And having visited once, would they not return now and then, to find out how the garden is growing?

In fact, reports of such visits appear regularly in the press. However, my feeling about these reports is one of

Contents

Introduction

If intelligent beings exist in space, and they look in our direction now and then, they will have observed something in recent years that must pique their curiosity. At television and FM frequencies, our planet blazes more brightly in the sky than the sun itself. For the first time in the history of the Universe, the earth has become a conspicuous object in the heavens—a beacon sending signals to every nearby star that intelligent life has arisen on this planet.

Some astronomers believe no intelligent beings exist on these stars to receive our message. They believe mankind is alone in the Universe. But other astronomers are unwilling to accept that conclusion. They do not find it credible that, out of all the billions of planets in the Universe around us, only one should have been touched with the magic of life.

These astronomers point to the fact that the Universe is 15 billion years old, but the sun and the earth are only four and a half billion years old. It follows that if other solar systems contain intelligent life, this life must, on the average, have evolved at least a billion years beyond mankind.

Reflecting on the fact that the human brain has doubled in size in the last million years, the astronomers say to themselves: Beings whose evolution has carried them a billion years beyond the human level may possess great heights of brainpower; how interesting it would be to speak to them, and find out what they have learned in that long interval. The greatest thinkers in human history might be as children in their presence. "In their eyes," one observer suggests, "Einstein would qualify as a waiter and Thomas Jefferson as a busboy."

How long must we wait before we are admitted to the company of these stimulating creatures? If the extraterrestrials exist, our television signals have made them aware of our presence; but perhaps they find our thoughts too simple to warrant a response. Perhaps another billion years must pass before the mature beings of the Cosmos judge mankind's descendants worthy of their attention.

It seems a pity that we must wait so long for the interesting exchange to begin. But this reasoning may underestimate the curiosity the advanced extraterrestrials might display in such new arrivals as Homo sapiens. Intelligence and curiosity go together; the discovery of new worlds may be life's greatest pleasure for a highly intelligent being. It is difficult to avoid a suspicion that the advanced beings—if they exist—have visited the solar system at least once in all those billions of years. And having visited once, would they not return now and then, to find out how the garden is growing?

In fact, reports of such visits appear regularly in the press. However, my feeling about these reports is one of doubt, because the beings that appear in UFO reports usually seem too humanlike to match a scientist's expectations for a form of life so far beyond mankind that it can cross the void between the stars. The extraterrestrial visitors are nearly always described as "humanoid." They differ from humans in size and detail, but always have the same basic design—made of flesh and blood, and with four limbs, two grasping appendages and a talking head.

Yet forms of life like this appeared on the earth only a few hundred million years ago. In the perspective of the 15-billion-year history of the Cosmos, that is a short time. It is possible that extraterrestrials visiting our planet will happen to be within a few hundred million years of the human level of development, but that is not very likely. Most life in the Universe—if other life exists—is either several billion years beyond us in its evolutionary state, or several billion years behind us. Such creatures, removed from mankind by billions of years of evolution, will not resemble us in any way.

The history of life on the earth provides a hint of the changes evolution works in the forms of life during a billion years. Life began on our planet about four billion years ago. A billion years later, it had evolved to the stage of bacteria and other one-celled organisms. Two billion years after that—and now the time is one billion years ago—evolution had reached the level of soft-bodied animals resembling the worm and the jellyfish. These creatures were the highest forms of life on the earth a billion years ago. In the last billion years of evolution, the fossil record tells us, the wormlike animals evolved into humans.

With this knowledge, derived from the fossil record and from the astronomical evidence for the antiquity of the Universe, what can we expect for the nature of life in other solar systems? If that life is a billion years behind mankind, it is likely to be at the level of the wormlike creatures which represented the most advanced organisms on the earth one billion years ago. In that case, this life could not have mastered the technology needed for the journey to our solar system. And if the life on another world is a billion years older, it is as far beyond us as we are beyond the worm—so far evolved beyond mankind as to bear no relation to humans in body, mind or desires. This line of thought leads me to the conclusion that extra-terrestrials arriving at the earth from another star are likely to be so different from the kind of life we know that we may not recognize them as living creatures when we see them.

This is the perspective, then, that science affords on mankind's place in the Cosmos: While we stand at the summit of creation on the earth, in the cosmic community of intelligent life we are among the children of the Universe. As we take our first steps into space, and reach out to cosmic life through our television and radar signals, it is my sense that the most momentous developments in the history of life on this planet may be about to unfold.

I
Dawn
of the
Space Age

THE ANCIENT SURFACE OF THE MOON. The multitude of craters in this NASA photograph conveys the antiquity of the lunar surface. The lunar landscape is scarred by countless craters formed in collisions with meteorites during past aeons. The earth has suffered a similar bombardment by meteorites during its history, but the surface signs of all but the most recent collisions have been eroded away by wind or water or buried by floods of lava. The moon preserves the record of its past better than the earth; events that occurred there billions of years ago are as freshly preserved as if they had happened yesterday.

HAROLD UREY (1898–1981) *(opposite)* was the father of modern lunar science and one of the greatest native-born scientists in American history. He received the Nobel Prize for his discovery of heavy hydrogen, or deuterium, made a major contribution to current ideas on the evolution of life out of nonliving matter, and invented an ingenious method for determining changes in the earth's climate that occurred hundreds of thousands of years ago.

I asked Dr. Urey once how he explained his ability to make four momentous contributions to science, when most scientists count themselves fortunate to make a single important contribution in their lifetimes. He replied, "I think it is because I only work on the big problems. I do not spend my time analyzing the calcium in the creek flowing past your door."

A SLIVER OF THE MOON. Dr. Urey *(below)* compares a tiny sliver of moon rock, lying on the pad in front of him, with two larger samples of volcanic rock from the earth. The moon rock was picked up and carried back to the earth by the *Apollo 14* astronauts. It comes from the edge of the mountainous lunar highlands.

The highlands are the oldest part of the moon, and date back to the beginning of the solar system, four and a half billion years ago. When this bit of rock was laid down on the moon, life was just beginning to evolve on the earth. The earth rock directly in front of Dr. Urey is only 180 million years old.

Harold Clayton Urey

ROCKS FROM THE MOON. Neil Armstrong was the first man to set foot on the moon. His copilot, Edwin E. Aldrin, followed him down the spacecraft's ladder 15 minutes later. The photograph of Aldrin *(opposite)*, perhaps the most famous photograph in the history of the space program, was taken by Armstrong. The reflection of his figure can be seen in Aldrin's visor.

Armstrong and Aldrin collected 842 pounds of rocks on the moon. When the rocks were brought back to NASA laboratories and their ages were measured, they turned out to be several billion years old, as Dr. Urey had predicted. Some bits of moon rock dated back to the very beginning of the moon's existence.

In the final moments of their descent to the moon, the two men narrowly escaped disaster. As the craft dropped toward the moon's surface under automatic pilot, they realized they were headed for a field of huge rocks, big enough to break apart their ship. Armstrong took over manual control and shot the landing craft forward over the moon's surface at nearly 60 miles an hour, searching desperately for a smooth landing site.

Armstrong found a small clearing, perhaps 50 feet across, headed for it, but then lost sight of the clearing in the clouds of dust raised by the engine of the descending craft. Feeling his way, he brought the spaceship down onto the surface. When it finally landed, only 20 seconds' worth of fuel remained; *Apollo 11* was seconds away from an aborted mission.

A LUNAR AUTOMOBILE *(below)*. In the last three moon landings, the explorers were aided by a lunar automobile—a lightweight car with an electric motor on each wheel and wire mesh "tires." The lunar automobile had a radio homing system that always told it where its mother ship was. The car weighed only 76 pounds in the moon's weak gravity. In *Apollo 17*, the last lunar landing, it carried the astronauts as far as 20 miles from their landing site and set a speed record for the moon of 11 miles an hour coming down off a mountain.

2

New Eyes
in
the Heavens

Midnight, July 20, 1969; a chiaroscuro of harsh contrasts appears on the television screen. One of the shadows moves. It is the leg of astronaut Edwin Aldrin, photographed by Neil Armstrong. Men are walking on the moon. We watch spellbound. The earth watches.

Seven hundred million people were riveted to their radios and television screens on that July night in 1969. What can you do with the moon? No one knew. Still, a feeling in the gut told us that this was the greatest moment in the history of life since the fishes left the water 350 million years ago. We were leaving the planet. Our feet had stirred the dust of an alien world. America had brightened its tarnished image. That was Apollo.

My fellow scientists never liked the Apollo project. In fact, they were never enthusiastic about any kind of manned space flight, and much preferred sending instruments into orbit to having people up there jiggling the telescopes. A

Nobel laureate in physics said once that notions about people traveling in space should "go back on top of a cereal box." The editors of *Nature*, the most prestigious science journal in the world, wrote a few years ago that manned space flight is "decades of research on why astronauts vomit."

I have disagreed with my colleagues on this issue for many years. When I was chairman of NASA's Lunar Exploration Committee, years before the lunar landings, the other scientists on the committee complained bitterly about the agency's plan for a manned flight around the moon, which was supposed to be a preliminary to the actual moon landing. A loop around the moon had no scientific value, they said.

Yet that flight around the moon, with its stirring Christmas Eve reading of the Book of Genesis from the moon by Frank Borman, Jim Lovell, and Bill Anders, was one of the greatest moments in human history.

I think my colleagues missed the point in those early days. NASA does very interesting science in space, but science has never been the only thrust of the space program. When Secretary of Defense Robert McNamara and NASA head James Webb wrote to President Kennedy, urging him to come out in support of a moon landing, they said, "It is man, not merely machines, in space that captures the imagination of the world."

Political leaders know that by instinct; scientists have to learn it the hard way. When I was a young space scientist and had just joined NASA, I thought, along with my colleagues, that the reason for going to the moon was to study the ancient moon rocks and find out what happened in the early years of the solar system. The memo Harold Urey and I wrote on why NASA should explore the moon conveyed our innocence. It said, "A soft landing of scien-

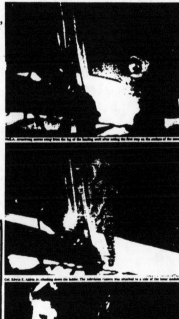

MEN ON THE MOON. The first landing on the moon commanded one of the largest headlines ever printed by *The New York Times*.

tific instruments on the moon will capture the imagination of the scientific community and the general public to a greater degree than any project of comparable scientific value."

As matters turned out, there was an error in our reasoning. We accomplished our goal; we landed elaborate packages of instruments on the moon, starting in 1964—but the feat did not have the impact on world opinion that Harold and I had expected. That came only later, when NASA landed *men* on the moon. Then the United States finally made its point.

The Apollo project cost about $75 billion in 1989 dollars. No one would spend that kind of money to satisfy a scientist's curiosity about how the earth and moon were formed. The main goal of the Apollo landing was not to bring back the moon rocks, but to develop a U.S. capability for operating and maneuvering in space, so that we could maintain our presence there in the face of vigorous activity on the part of another great power. Some instinct told the American people that the mastery of space flight would be vital to the national interest.

The strong public interest in manned space flight continued beyond the lunar landings and into the early days of the Space Shuttle. The Shuttle was a reusable rocket, designed to be less expensive than the old-fashioned kinds of rockets used in the Apollo program, that were fired once and then thrown away. But the Shuttle needed a crew of astronauts to bring it back to earth. Making the Shuttle "man-rated"—acceptably safe for its astronaut crew—added considerably to its cost, and offset the savings gained by the Shuttle's reusability.

To recapture a low cost per flight, NASA argued for and won the assignment to use the Shuttle for all the nation's business in space—not only activities that required human participation, but the launching of ordinary, work-aday unmanned satellites as well. The idea was that the unit cost of launching Shuttles into orbit would be cut by a high-volume operation.

But a rapid turnaround of the Shuttle fleet on a routine basis, with flights every few weeks, turned out to be much more difficult than NASA had expected. The Space Agency found itself unable to meet the flight schedule that had been promised when the Shuttle was in its planning stage. Pressure built up within NASA to preserve a lively Shuttle launch schedule in spite of the technical difficulties. These circumstances set the stage for the Challenger accident.

The immediate cause of the accident was a failure of the seals that keep hot gases inside the burning rocket from leaking out. When these hot gases came into contact with the exterior of the rocket, they weakened its structure and the Challenger rocket broke apart and exploded.

But the deeper cause of the accident was the pressure on NASA to maintain a fast-paced schedule of Shuttle launches. That fast-paced schedule was the product of the unfortunate earlier decision to have the Shuttle carry the burden of launching every U.S. satellite—a decision driven, in turn, by the need, imposed on NASA by the Administration and Congress, to justify the Shuttle as a cost-saving enterprise.

Even before the Challenger accident, space scientists did not look kindly on the Shuttle because they felt it was drawing scarce funds away from their purely scientific work in space. *Nature* spoke for many in the scientific community when its editors complained that the Shuttle ate up the "lion's share" of the NASA budget. In the early days of NASA, that did not matter very much; while all those billions were pouring into the NASA budget for the trip to the moon, another $50 million or so was always available for a scientific satellite. But today, things are different. You cannot buy much science in space for $50 million any more. The spectacular Voyager flight to the planet Jupiter and its moons—one of the greatest triumphs of planetary exploration thus far—cost more than $400 million in 1989 dollars. The Viking spaceship that landed on Mars and searched for life cost nearly two billion dollars. The four large telescopes NASA plans to have in orbit during the 1990s range in cost from more than $500 million to well over two billion dollars each.

As the cost of the biggest scientific projects in NASA went up, the NASA budget went down. In the halcyon days of the Apollo program, during the 1960s, NASA received more than $20 billion every year (in 1989 dollars) for several years running. But in the 1970s, the NASA budget declined to under six billion dollars a year. No managerial or technical wizardry will buy a twenty-billion-dollar space program for six billion dollars.*

In this new era of billion-dollar orbiting observatories and smaller NASA budgets, astronomers and planetary scientists have not fared as well as they once did in the competition with man-in-space. But now science in space— and especially astronomy—is beginning to get some important returns from the manned space program. While the big orbiting observatories are designed to work in space for years, they may not live out their allotted life spans;

---

*During the same period, the U.S. economy grew from two trillion dollars (in 1989 dollars) in the mid-1960s, to five trillion dollars today. If the NASA budget today were as large a fraction of the U.S. economy as it was during the Apollo program, it would be roughly $50 billion a year.

even if their instruments continue to function perfectly, a meteorite or a bit of man-made space debris may damage a mirror or puncture a vital electronic circuit. Thousands of bits and pieces larger than a few inches across—including a camera and a screwdriver—circle the earth in orbit, left over from earlier space missions. In a collision at satellite speeds, hurtling through space at five miles per second, each chunk of debris has the explosive power of a hand grenade.

Two of NASA's space observatories—the Hubble Space Telescope and the Advanced X-ray Observatory—cost considerably more than all the funds spent by U.S. scientists for astronomy on the ground in the last twenty years. To make certain the NASA observatories last into the next century in spite of the hazards of space, they have been designed so that they can be repaired in orbit by astronauts and scientists ferried up on the Shuttle or working out of the Space Station. These great space observatories are too expensive to be abandoned if they break down in orbit or collide with space debris. If they stop working, astronomy will suffer an incalculable loss because they probably will not be replaced in our lifetime.

•

Putting telescopes into orbit is perhaps the most valuable single contribution the space program can make to astronomy. Because a telescope in space is far above the obscuring effect of the earth's atmosphere, it can produce sharper and clearer images than the finest telescopes on the ground—more than ten times as sharp as the images produced by the 200-inch telescope on Palomar Mountain. Space telescopes may provide answers to some of the greatest mysteries in the Cosmos: When did the Universe begin? What forces brought it into being? How large is the observable Universe?

How can a telescope provide information about the beginning of the Universe? The answer is that when we look out into space, we look into the past. If a galaxy is five billion light-years away, it takes five billion years for the light from this galaxy to reach the earth. Consequently, our telescopes show the galaxy not as it is today, but as it was five billion years ago, when the light we are receiving now had just left that galaxy on its way to the earth. A telescope is a time machine; it carries us back into the past. Telescopes in space, which will see farther out into the Universe than any telescope on the ground, will also see farther back into the past. They may tell us whose hand was at work in the moment of creation.

The telescopes that have been sent into space or will be going into space are fairly modest in size; the largest is barely half as big as the great 200-inch telescope on Mount Palomar. Why do these relatively small telescopes in space see farther than the 200-inch telescope on the ground?

The answer is connected with the fact that rays of light from stars and galaxies have to pass through the earth's atmosphere to reach a telescope on the ground. But the atmosphere bends the light rays, just as glass bends a ray of light. The amount of bending varies from moment to moment, which causes the light rays to shift erratically. This blurs the image formed by the telescope. The astronomer on the ground, trying to see the heavens through the earth's atmosphere, is like a person peering at the world through a poorly made window.

The blurring of the telescope's images limits the

distance of the celestial objects it can detect. Consider a very distant object—a galaxy, for example. This distant galaxy produces a faint image in the telescope, because the intensity of the light from any object becomes weaker with inceasing distance. Now, if that faint image is spread out and diffused by the blurring effect of the earth's atmosphere, it becomes fainter still. In fact, it tends to be submerged in the background illumination of the night sky, so that it cannot be seen at all.

If the astronomer's telescope is in space, the blurring effect of the atmosphere is eliminated, the images formed by the telescope are sharper, and they stand out better against the background sky illumination. That is why telescopes in space can see farther out into the Universe, and farther back in time.

Telescopes in space have other advantages. Nearly all the information we have about the Cosmos comes to us in the form of radiation of various wavelengths, which passes through the atmosphere to the surface of the earth. But most of this radiation is absorbed by the atmosphere before it reaches the surface. Gamma rays, X rays, and most wavelengths of infrared radiation are blocked; the only part that gets through is the narrow band of wavelengths called "light" or visible radiation (because the human eye is sensitive to this particular band of wavelengths), as well as some radio waves and a bit of infrared radiation.

But some of the most interesting objects in the Universe, such as black holes and quasars, emit copious amounts of just those kinds of radiation that cannot get to the ground—particularly gamma rays and X rays. And stars with planets forming around them emit copious amounts of infrared radiation, another kind of radiation blocked by the atmosphere. Few objects in the Cosmos are more interesting than black holes, quasars, and solar systems in formation, yet most of the information about these exotic, important objects was denied to astronomers before the advent of the space age. When NASA started to put gamma ray, X-ray, and infrared telescopes in satellites orbiting above the earth's atmosphere, new windows opened on the heavens.

The Gamma Ray Observatory, one of the new eyes in space designed to look through those windows, will be in some ways the most interesting astronomical satellite of all. Gamma rays have tremendous penetrating power in space and can travel great distances through the Universe. When we see a gamma ray that has come from a distant place, we see a different view of the Universe, as it was at a very early time, perhaps not long after its birth.

The picture of the early Universe afforded by the Gamma Ray Observatory may provide clues to another of the great cosmic mysteries: Where is the antimatter in the Universe? Physicists have discovered that two kinds of matter exist, with properties that are similar, but, in some senses, opposite. They call these "matter" and "antimatter." For example, the electron, one of the building blocks of ordinary matter, has an antimatter counterpart called the antielectron, or positron. The electron and its antimatter counterpart, the positron, have the same mass and are almost identical, except that they carry electric charges of opposite signs—a negative charge for the electron and a positive charge for the antielectron, or positron.

In the same way, the proton, a building block of ordinary matter, has an antimatter counterpart called the antiproton. While the proton carries a positive electric charge, the antiproton bears a negative electric charge.

Other experiments performed by physicists have revealed the remarkable fact that when matter and antimatter come into contact, they annihilate each other, leaving behind only radiant energy in the form of gamma rays. For example, when an electron and an antielectron meet, the two particles disappear, and a flash of gamma radiation emerges from the point at which they met.

But if two particles of antimatter meet, they do not annihilate each other. In fact, antimatter particles can be brought together and combined to make an atom of antimatter, and antimatter atoms can be combined into entire pieces of antimatter. For example, just as a hydrogen atom consists of an electron circling around a positively charged proton, an anti-hydrogen atom can be formed from an antielectron circling in orbit around an antiproton. And, of course, antimatter equivalents of all the other elements— anti-carbon atoms, anti-oxygen atoms, and so on—can be built as well.

In other words, according to everything known about the laws of physics, an entire world of antimatter can exist. In the antimatter world there can be antimatter galaxies with antimatter stars, antimatter planets, and antimatter life.

How interesting it would be to discover antimatter stars inhabited by intelligent antimatter beings. We could never make contact with them, of course; the meeting would annihilate the material substances of both participants, leaving behind only gamma rays.

But scientists have found no evidence for antimatter planets, stars, or galaxies. And so they ask themselves: Where is the antimatter in the Universe?

The Gamma Ray Observatory may be able to tell us. Since collisions between galaxies are fairly common, if antigalaxies exist, they must collide now and then with ordinary galaxies. When such a collision occurs, part of the material in the galaxy and the antigalaxy will be annihilated in a shower of gamma rays. If these sparks of annihilation exist, the Gamma Ray Observatory should be able to detect them and test this fascinating hypothesis.

Finally, there is the question of the intense bursts of gamma rays—the bursters—that seem to blast out of the heavens occasionally. Often they occur in parts of the sky where no galaxy, star, or other object can be seen. What are they? What massive object can produce gamma rays without, at the same time, producing visible light, ultraviolet radiation, and other kinds of radiation? No one knows the answer to that question either. The Gamma Ray Observatory may solve the mystery of the gamma ray bursters.

Another astronomical observatory in space will scan the heavens for infrared radiation, or heat. Infrared radiation has been an invaluable source of information for astronomers. Several years ago, a satellite carrying an infrared telescope discovered evidence that planets were forming around several stars in the sun's neighborhood. This was the first clear indication that solar systems like ours are commonplace in the Universe—a portentous discovery for the prospects of extraterrestrial life. That early infrared satellite also discovered six new comets in our solar system, as well as the nucleus of a dead comet that could collide with the earth, according to the calculations of one astronomer. If the collision occurs, it will release the energy of 200,000 hydrogen bombs, with catastrophic results for life on the earth.

All told, four great astronomical observatories are scheduled to be launched in this decade and the next. These billion-dollar observatories are the jewels in the crown of space science. They promise a revolution in astronomy as great as the revolution that occurred in 1609, when Galileo first raised the newly invented telescope to the heavens to discover the phases of Venus and the moons of Jupiter—and the era of naked-eye astronomy drew to a close.

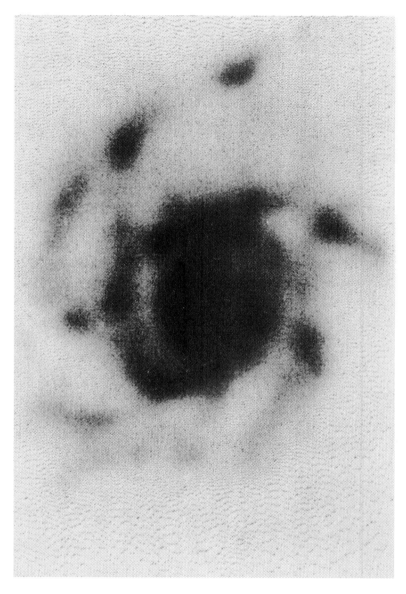

SHARPER IMAGES IN SPACE. A comparison of the illustrations
on this and the facing page shows how the images formed by a
telescope in space are sharpened by its location above the atmosphere. The photograph shows a distant galaxy as it might appear
through the 200-inch telescope on Palomar Mountain—one of the
world's largest. Details are blurred by the shifting motion of rays of
light from the galaxy as they pass through the earth's atmosphere
on their way to its surface.

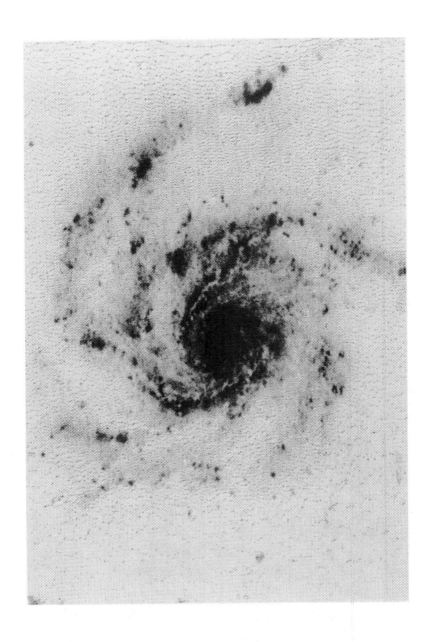

This illustration shows the same galaxy as it would appear photographed from space through the Hubble Space Telescope. The image is 20 times sharper. The blurring effect is avoided because the telescope is in a satellite above the atmosphere.

SPACE DEBRIS: A HAZARD TO SCIENTIFIC SATELLITES. Debris in space is a growing hazard for scientific satellites and manned spacecraft. A collision with a tiny flake of paint created a one-tenth inch hole in the skin of a NASA satellite *(below)*. Another flake of paint gouged out a quarter-inch pit in the windshield of the space shuttle *Challenger*. The energy of the impact with a flake traveling at three miles a second was equivalent to a bowling ball hitting the windshield at 60 miles an hour.

The illustration *opposite* shows the instantaneous positions of some of the more than 7,000 pieces of man-made debris—dead satellites, empty rocket casings, and fragments of destroyed space-craft—that are orbiting the earth and are big enough to be detected and tracked by radar. The debris being tracked includes an astronaut's glove and camera, and a Soviet Cosmonaut's screwdriver. A collision with any of these thousands of objects would probably destroy the Hubble Space Telescope.

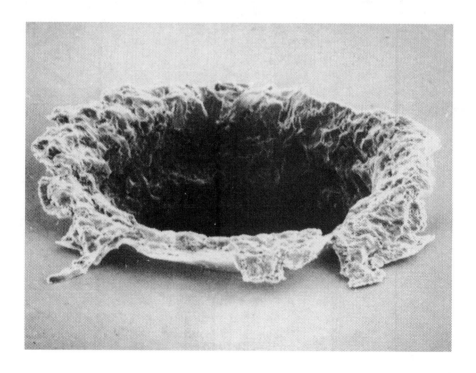

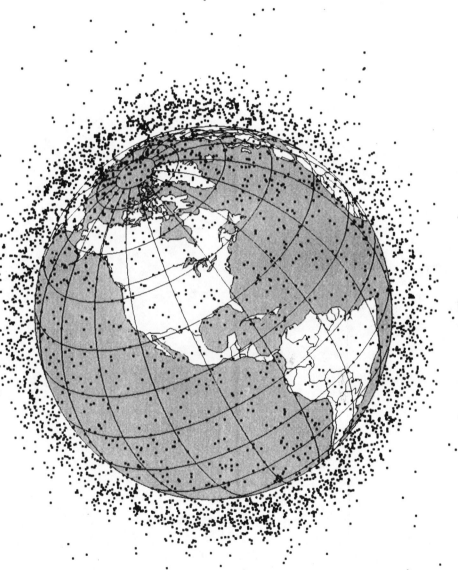

II
The Cosmic
Mysteries

3

In the
Beginning

Dramatic discoveries in recent years have transformed man's picture of the Universe. No longer is the Cosmos a quiet place, traversed by stars and planets moving in stately procession. Today we know it to be richly populated by exotic objects, lashed by savage forces, and pregnant with surprises. Strange quasars light up the dark corners of the Universe; massive galaxies race across the sky; titanic explosions of unknown origin occur in the depths of space; and tentative evidence has been uncovered for the black hole—the most bizarre object ever conceived by the scientific mind.

Most remarkable of all, astronomers have found proof that the Universe sprang into existence abruptly, in a sudden moment of creation, as the Bible said it did. And they have come upon evidence that man's birthplace was in the stars. By small increments of knowledge, accumulating slowly over five centuries, they have pushed the frontiers of the material Universe further and further back into the world of what used to be considered the spiritual. The result of their efforts is a chronicle of events by which the simple atoms in the primordial cloud of the Universe gradually evolve into conscious life. It is a story of continuously unfolding creation—some say the continuous unfolding of some great plan or design, but others say there is no plan.

I myself do not know what to think. I am an agnostic in these matters. But the story itself is very interesting. It can be told simply, without mathematics or jargon. Let us begin at the beginning.

In 1912, an American astronomer named Vesto Melvin Slipher decided to turn his telescope on the great clusters of stars called galaxies. Most stars are clustered in groups called galaxies, just as people are clustered together in nations. These galaxies drift through space like island universes, with large distances between them. Our sun belongs to the cluster of 200 billion stars called the Milky Way Galaxy, which has the shape of a giant, flat spiral, nearly a million trillion miles in diameter.

Half the galaxies in the Universe are flat, rotating spirals similar to ours. Our nearest large galactic neighbor, the Andromeda Galaxy, is also a spiral galaxy, containing another 200 billion stars.

The Andromeda Galaxy is 12 million trillion miles away. To avoid the frequent repetition of such awkwardly large numbers, astronomical distances are usually expressed in terms of the light-year, which is the distance covered in one year by a ray of light traveling at a speed of 186,000 miles per second. In the astronomer's units, the distance to Andromeda is two million light-years. Thousands of galaxies exist within a distance of 100 million light-years away from us, and many billions more are within the range of the largest telescopes.

Now we come to the extraordinary discovery that lies at the heart of the scientific theory of creation. In 1913, in the course of his study of galaxies, Slipher found that most galaxies within range of his telescope were moving through space at extraordinarily high speeds, in some cases as great as several million miles an hour. Furthermore, nearly all these galaxies were moving away from the earth.

According to astronomical knowledge, our planet and its parent star, the sun, are indistinguishable from countless other planets and stars in the heavens. Why should all the galaxies in the Universe move away from us? We would expect them to move randomly, so that at any moment half the galaxies in the Universe would be moving toward us and half would be moving away. But according to Slipher's measurements, this was not so; the entire Universe was moving away from one special point in space, and the earth was located at that point.

That would seem to imply, in turn, that the earth is at the center of the Universe. Very few people accept that notion today. Why does modern astronomy lead to a picture of the Universe that was abandoned by scientists many years ago?

The answer is that Slipher's findings only seem to lead to the conclusion that the earth is at the center of the Universe. If you were in another galaxy in the Universe, you would see all the galaxies around you moving away, in exactly the same way that we see all the galaxies around us moving away. You would think that you are at the center of the expanding Universe, and so would every astronomer in every other galaxy; but, in fact, there is no center.

To understand this statement more clearly, imagine a loaf of raisin bread baking in the oven. Each raisin is a galaxy. As the dough rises in the oven, the interior of the loaf expands, and all the raisins move apart from one another. The loaf of bread is like our expanding Universe. Every raisin sees its neighbors receding from it; every raisin seems to be at the center of the expansion; but there is no center.

Of course, to make the analogy more precise, we would have to imagine a loaf of raisin bread so large that you could not see the edge from the interior, no matter where you were located; that is, the loaf of bread, like the Universe, would be infinite.

In the 1920s and 1930s, astronomers Edwin Hubble and Milton Humason, using the 100-inch Mount Wilson telescope, then the largest telescope in the world, succeeded in measuring the speeds of many other spiral galaxies. These galaxies were too faint to have been seen by Slipher with his smaller instrument. The observations by Hubble and Humason confirmed Slipher's discovery; without exception, all the distant galaxies in the heavens are moving away from us and from one another. The entire Universe appears to be exploding.

The picture of an exploding Universe has a remarkable implication. If the galaxies are moving apart today at high speeds, at an earlier time they must have been closer than they are now. At a still earlier time, they must have been closer yet. If we continue to trace the motions of the outward-moving galaxies backward in time, we find that they all come together, so to speak, 15 billion years ago. At that time, all the matter in the Universe was packed into a dense, hot mass, with temperatures ranging up to trillions of degrees. The picture suggests the conditions in an exploding hydrogen bomb. The instant in which the cosmic bomb exploded marked the birth of the Universe.

The idea that the Universe exploded into being, which rests on Slipher's discovery of the motions of the galaxies, is often called the *Big Bang* theory, or Big Bang cosmology. The seed of everything that has happened since in the Universe was planted in the first instant of the Big Bang; every star, every planet, and every living creature in the Cosmos can trace its substance back to the moment of the cosmic explosion. It was literally the moment of creation.

This is a curiously biblical view of the origin of the world. The details of the astronomer's story differ greatly from those in the Bible; in particular, the age of the Universe appears to be far greater than the 6,000 years of the biblical account; but the astronomical and biblical accounts of Genesis are alike in one essential respect. There was a beginning, and all things in the Universe can be traced back to it.

The exact moment in which the Big Bang occurred is uncertain by some billions of years. Because of this uncertainty, I have picked 15 billion years, a round number, as *the* age of the Universe; but the correct age could be as little as 10 billion years or as much as 20 billion years. The important point is not precisely when the cosmic explosion occurred, but that it occurred billions of years ago, at a sharply defined instant.

Many astronomers have found the idea of an abrupt beginning philosophically distasteful, even if the beginning occurred as long ago as 15 billion years. They have been reluctant to accept the notion that the Universe has not existed forever. However, most skeptics were convinced some years ago, when a second great discovery was made. In 1965, two physicists, Arno Penzias and Robert

Wilson, detected a puzzling glow of radiation coming from the sky. The radiation was very faint and diffuse and did not seem to come from any particular object, such as the sun or the moon. The entire Universe seemed to be the source of the strange radiation. Penzias and Wilson were mystified.

But other astronomers, who had been studying the evidence for the Big Bang—the explosive birth of the Universe—knew immediately what this radiation was. Their calculations had shown that if the world really exploded into existence billions of years ago, the Universe would have been filled with a fireball of white-hot matter and radiation immediately after the explosion. As the Universe expanded and cooled, the light of the fireball would have become less intense, but a remnant of the cosmic fireball would still be detectable today. It was this echo of the cosmic explosion that the two physicists had detected; they had stumbled on the proof that the Universe had a beginning.

These results seemed quite strange to most astronomers, and they have tried very hard to find other explanations for the radiation discovered by Penzias and Wilson —explanations that would not involve the idea of an abrupt beginning of the world. However, still another finding indicates that these other explanations do not work; only the cosmic explosion can explain the puzzling radiation.

The new finding is that the wavelengths, or colors, present in this radiation almost exactly match the pattern of wavelengths expected for the light and heat produced in an explosion. This pattern of wavelengths is the fingerprint of the Big Bang, blurred after billions of years, but still unmistakable. The match between the pattern of wavelengths in an explosion and the pattern of wavelengths in the fireball radiation has convinced nearly all scientists that the Big Bang really did occur.

However, that picture creates a serious problem for science. Scientists are people who believe every event that occurs in the world can be explained by science in a rational way as the product of some previous event; they believe that science has an answer to nearly every question. Therefore, they ask themselves, "Why did the Universe begin in an explosion? What did the Universe look like before the explosion? Did the Universe even exist prior to that moment?"

To these questions they can find no answer—at least, not in science. James Peebles, an American astronomer who has made important contributions to the theory of the expanding Universe, said some years ago, "What the Universe was like at day minus one before the Big Bang, one has no idea. The equations refuse to tell us. I refuse to speculate." And the British physicist, Edward Milne, reflecting on the evidence for the expanding Universe, wrote, "We can make no propositions about the state of affairs in the beginning; in the Divine Act of creation God is unobserved and unwitnessed."

GALAXIES *(opposite)*. Stars are clustered together in galaxies, just as individuals are clustered together in nations. The galaxies are scattered through the Universe and separated by largely empty space. Each galaxy is a cluster of billions of stars, held together by the force of gravity.

Many galaxies rotate, and are flattened into the shape of a disc or plate by their spinning motion. The photograph *opposite* shows a spinning galaxy, listed in astronomical catalogues as NGC 3031. The many billions of stars in this galaxy are so closely packed that they present the appearance of a continuous mass of luminous matter.

Our sun belongs to a flattened, rotating galaxy called the Milky Way, which looks like NGC 3031. If we could step outside our galaxy and view it edge on, it would look like NGC 4565 *(below)*.

NGC 4565: A rotating galaxy like NGC 3031, seen edge on.

NGC 3031: A rotating spiral galaxy (opposite).

PENZIAS AND WILSON. Arno Penzias and Robert Wilson *(opposite)* received the Nobel Prize in 1978 for their detection of the primordial fireball radiation—one of the greatest scientific discoveries of all time.

The chart *below* shows the results of measurements considered by astronomers to be the clinching evidence that the Universe began in an explosion. The shaded area on the chart indicates how the intensity of the cosmic fireball radiation varies with wavelength. (The width of the shading represents the uncertainty in the measurements.) The chart shows that the radiation is most intense at a wavelength of 1.8 millimeters, or one fiftieth of an inch, and falls off to weak intensities at shorter and longer wavelengths.

The solid line shows the variation of intensity with wavelength for the radiation produced by an explosion. The agreement between the shaded area and the line has convinced most astronomers that the Universe had an explosive beginning.

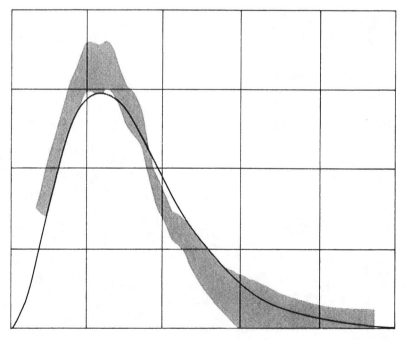

Variation of intensity with wavelength in the cosmic fireball radiation. The shaded area encompasses the measurements; the solid line is the theoretical curve for the energy radiated in an explosion.

Arno Penzias *(left)* and Robert Wilson receive the 1978 Nobel Prize for Physics in Stockholm.

4

Birth of
the Stars

It is possible that astronomers will never discover an explanation for the beginning of the Universe. But while science cannot explain what came before the beginning, it offers a good account of the events that came after. With the aid of telescopes and other instruments, scientists have succeeded in reconstructing the chain of events by which the gases of the newly created Universe were gradually transformed into stars, planets, and life.

The narrative pieced together by scientists starts immediately after the Big Bang, when the Universe was very dense and very hot. The very elements that make up the earth and its inhabitants did not yet exist. The Universe was a fiery sea of radiation, from which particles emerged only to fall back, disappearing and reappearing ceaselessly.

The hot Universe expanded rapidly, and when it was one second old, its density had fallen to the density of water and its temperature had decreased to a billion degrees. At this time, the fundamental building blocks of matter—electrons, protons, and neutrons—condensed out of the sea of hot radiation like droplets of molten steel condensing out of the metallic vapor in a furnace.

The Universe continued to expand, and the temperature dropped further. By the time the Universe was three minutes old, the temperature had fallen to around ten million degrees. At that point, protons and neutrons stuck together in groups of four to form helium nuclei.

After the first three minutes, nothing of importance happened for the next million years. A glow of radiation, left over from the cosmic fireball, pervaded the Universe, obscuring visibility like a thick fog. Particles moved erratically through the fog, colliding with other particles, and sometimes colliding with packets of radiant energy.

When the Universe was about one million years old, an important event occurred: the first atoms appeared. An atom consists of electrons circling in orbit around a nucleus. When the Universe was very young, it was also very hot, and collisions between particles were violent. As a consequence of these violent collisions, whenever an electron was captured into an orbit around a nucleus to form an atom, it was knocked out of its orbit almost immediately. But when the Universe reached an age of about a million years, its gases were sufficiently cool so that collisions were more gentle. Now many electrons could remain in orbit after they were captured. From this moment on, the matter in the Universe consisted mainly of atoms of hydrogen and helium, swirling through space in vast clouds.

At the same time the obscuring fog of radiation cleared up, and the Universe suddenly became transparent. The reason for this change was that light, which is a form of radiation, cannot pass through electrically charged particles. The electrons and protons that were moving around in space before the formation of atoms were, of course, electrically charged. An atom, on the other hand, is electrically neutral because the positive electric charge on the proton cancels the negative charge on the electron. Therefore, atoms do not block radiation appreciably.

As soon as the electrons in the Universe had combined with protons or other nuclei to form neutral atoms, rays of light were able to travel great distances unhindered, and it became possible to see from one end of the Universe to the other. But no eye was present then to perceive the Universe. Neither galaxies, stars, planets, nor life had yet appeared. They were still to come.

After the appearance of atoms, no other changes of consequence occurred for a long time, except that the Universe continued to expand, gradually becoming cooler and less dense. But when the Universe was about one billion years old, a dramatic sequence of events began.

Imagine the Universe at that early time, in existence for a billion years, filled with clouds of hydrogen and helium surging and eddying through space. No lights existed yet in the heavens. The Universe was a dark place. Now, in the mind's eye, see the swirling clouds, thin in some places, and dense in others. The pattern shifts quickly; the dense regions do not persist long; the atoms in them fly apart again as a consequence of their random motions, and the dense cloud quickly disperses to space.

But each atom exerts a small gravitational attraction on its neighbors. The attraction tends to pull the atoms together and counters the tendency of the cloud of atoms to fly apart. If the number of atoms in the cloud is very great, the sum of all these separate forces of attraction will keep it together indefinitely. It is then a tightly bound pocket of atoms, held together by the attraction of each atom to its neighbors. This cloud of gas in space, held together by gravity, is an embryonic star.

Once formed, the cloud exerts a gravitational pull on the material at its surface, just as the earth, or any other large body, exerts a pull on the objects at its surface. Under the influence of this pull, the atoms at the surface begin to "fall" toward the center of the cloud.

The atoms continue to fall; as they fall, they pick up speed, and their energy increases. The increase in energy heats the gas and raises its temperature. Gradually, under the pull of gravity, the cloud contracts. As it contracts, it grows hotter.

After about ten million years, the temperature at the center of the shrinking cloud rises to a critical value of 20 million degrees Fahrenheit. At this temperature, a nuclear reaction flares up in the center of the cloud, similar to the nuclear reaction that provides the energy for a hydrogen bomb. The reaction releases vast amounts of nuclear energy. The energy flows to the surface, causing the ball of gas to glow brilliantly. A new light has appeared in the heavens. A new star has been born.

The star lives on, burning nuclear fuel at its center, until the fuel is nearly exhausted. In the case of a star the size of the sun, that takes about ten billion years. Since the sun was born 4.5 billion years ago, it can be said to be a middle-aged star; it has half its life still before it.

As time passes and the star gradually ages, the nuclear fires in its interior burn steadily, consuming hydrogen and helium—the primordial stuff of the Universe—leaving behind a residue of heavier elements. These heavier elements are the ashes of the star's fire. Oxygen, iron, copper, and many other elements are included among the ashes. According to astronomers, all the elements in the Universe other than hydrogen and helium are the ashes of nuclear reactions in stars. Some elements are cooked in a star's interior during its long life. Others—the heaviest elements—are made in the brief moments of the catastrophic collapse and explosion that terminate the star's existence.

Toward the end of a star's life, when the reserves of nuclear fuel at the center are nearly depleted, the star begins to show signs of its age. The first sign of approaching demise in a star is a pronounced swelling and reddening of the star's outer regions. Such aging, swollen stars are called *red giants*. The sun will swell to a red giant in six billion years, engulfing the inner planets of the solar system in hot gases and vaporizing the earth and any creatures that may be left on its surface. Probably man's descendants will have fled the solar system by then, and gone to a younger star.

After a brief interval of some millions of years as a red giant, a star begins to collapse once more under the force of gravity. What happens next depends on how big the star is. Stars come in different sizes—small, medium, and large. If a star is modest in size, its collapse continues until it becomes a shrunken sphere of extremely dense, white-hot matter. These shrunken stars are called *white dwarfs*. A pint-sized container of matter from a white dwarf would weigh a thousand tons.

Slowly the white dwarf radiates the last of its heat into space and fades into oblivion. Eventually, it becomes a cold, dark lump of matter and enters the graveyard of the stars.

Surrounding the white dwarf is a distended shell of gaseous matter—the remains of the swollen red giant. These gaseous shells, glowing with myriad colors, are objects of extraordinary beauty. They are called *planetary nebulas* because the astronomers who first discovered them thought that they looked like families of planets—that is, new solar systems—in the process of forming out of gas and dust. We know today that the planetary nebulas have no connection with planets or solar systems, but the name has remained.

The sun, which is a moderately small star, will end its days in this way, as a white dwarf that casts off its outer shell of glowing gases and fades to yellow, then to red, and finally into invisibility.

A very different fate awaits massive stars, larger than the sun. Because these stars are so massive, their final collapse is a catastrophic event. Huge pressures converge on the core of the star, squeezing the material at its center into a dense lump. The outer layers of the star pile up around the dense lump at the center—and then the hot, compressed gases rebound in a violent explosion, blowing the outer layers of the star into space.

The final collapse and explosion of a massive star are cataclysmic. A glowing cloud of debris expands from the site of the explosion at a speed of thousands of miles per second, expanding to a radius of about a billion miles in one day. The exploding star is called a *supernova*. Supernovas are among the most spectacular events in the Universe. Every few hundred years, a supernova explodes in our own galaxy. When it does, we suddenly see a new star in the sky, brighter than a billion ordinary stars, and so bright that it may be visible in the daytime.

Europe witnessed a spectacular supernova explosion in 1572, and another in 1604. These supernovas were in our own galaxy, just a stone's throw from the sun in the scale of cosmic distances. For weeks they were brilliant beacons in the night sky, and then they faded into invisibility. At that time no one knew that a star is a sphere of flaming gas that forms in the heavens, lives for a time, and dies. That understanding would not come to science for another four centuries. Two thousand years earlier, Aristotle had claimed the heavens were unchanging and constructed his Universe on that assumption. Everyone believed he must be right; but now—two new stars—visible to all!

No supernovas exploded again in or near our galaxy until 1987. On February 23 of that year, a night assistant at the Las Campanas Observatory in Chile interrupted his labors to go outside for a walk. Looking up, he saw something peculiar in the Large Magellanic Cloud, a nearby and fairly small galaxy of a few billion stars, held captive to ours by the force of gravity. Knowing the stars of the southern sky like the back of his hand, the assistant now saw a star in a place in the Large Magellanic Cloud where no star had ever been seen before. The new star was a supernova—the first to occur close to our galaxy since the dawn of modern astronomy.

Supernova explosions have a special importance for the scientist interested in reconstructing the chain of events that led to the appearance of mankind on the earth. When a supernova explosion occurs, the body of the exploding star is sprayed into space. The material dispersed to space includes atoms of carbon, nitrogen, oxygen, and many other substances that were manufactured in the star's interior during its lifetime. In space, these atoms mingle with the fresh hydrogen and helium of the galaxy to form a gaseous mixture containing all the elements known in the Universe.

Later, other stars form out of clouds of hydrogen and helium that have been enriched by the products of many stellar explosions in the past. The sun is one of those stars; it contains the remains of countless stars that lived and died in the earlier years of the Universe. The planets also contain these stellar remains, and the earth and all life on the earth, including mankind, are formed almost entirely of them. We are all fashioned from the debris of dead stars.

BIRTH OF STARS. The dark spots scattered across the face of this photograph of the nebula known as NGC 6611 are pockets of dense gas and dust held together by gravity. Each pocket is a few hundred billion miles across. These pockets of gas are not yet stars, but they are on their way to becoming new stars. The glowing clouds are masses of gaseous matter that have been heated by the absorption of radiation from the hot, young stars that have recently formed in their midst.

THE CONE NEBULA: ANOTHER SEEDBED OF STARS. This region of our own galaxy, known to astronomers as the Cone Nebula, or NGC 2237, contains many clouds of gas and dust that are the seedbeds of new stars. Several hot, newly formed stars are visible in the photograph. The dark, blunt-tipped cone, extending upward toward the center, is a large and relatively dense cloud that blocks the light from the stars behind it. The brightest stars in this and the adjoining photograph are probably less than a million years old.

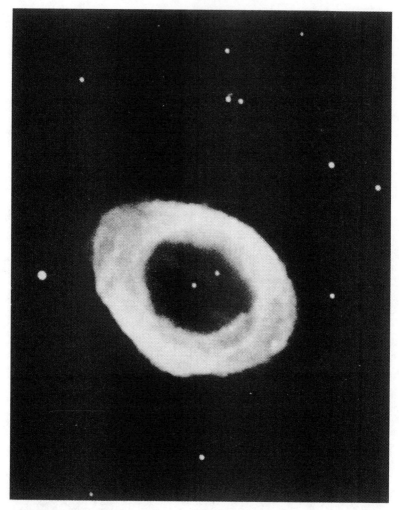

DEATH OF A SMALL STAR. This photograph shows a star of moderate size that has thrown off its outer region as an expanding shell of luminous gas. The dense core of the original star is visible in the center of the expanding shell. The sun will throw off its outer envelope in this way when it nears the end of its life in approximately five billion years.

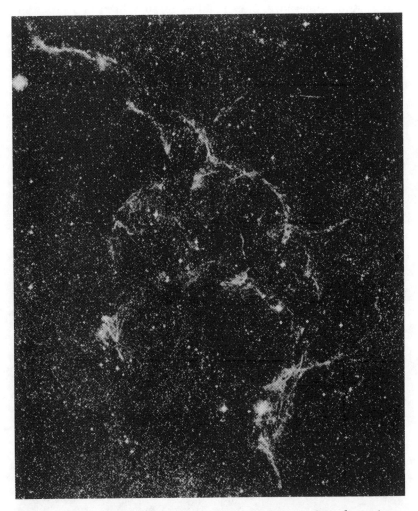

DEATH OF A MASSIVE STAR. Large stars end their lives in a titanic explosion called a supernova. The billowing cloud of gaseous matter *above* is the remnant of a massive star that came to the end of its life and exploded 10,000 years ago. The squeezed core of the exploded star is at the center of the small white circle. The cloud is at a distance from us of 1,500 light-years, or approximately ten thousand trillion miles.

65

ASTRONOMICAL EVENT OF THE CENTURY. On February 23, 1987, astronomers in the southern hemisphere observed the closest and brightest supernova, or exploding star, to be seen in nearly four centuries. The 1987 supernova occurred in a dwarf galaxy called the Large Magellanic Cloud *(right)*, which contains a few billion stars. The Large Magellanic Cloud is a satellite galaxy held captive by our galaxy's gravity, as the moon is held captive by the earth's gravity.

The supernova occurred in a spidery luminosity known as the Tarantula Nebula near the top of the Large Magellanic Cloud. The Tarantula Nebula is another breeding ground of stars that contains many massive, hot young stars. The photographs *below* show the sky near the Tarantula Nebula before and after the supernova exploded.

The star that exploded was 20 times as massive as the sun and had lived for ten million years. In its youth it was considerably hotter than the sun and 50,000 times brighter. Blue in color originally, it turned red, and then blue again in the last years of its life, evolving at an accelerating pace at the end.

In rapid succession, nuclear fires burning at the center of the star produced carbon, oxygen, neon, magnesium, sulphur, silicon, iron, and other elements. In a matter of seconds, the star collapsed and rebounded in the explosion observed by terrestrial astronomers.

During the first second of the explosion, the energy emitted by the dying star was greater than the energy emitted by the entire visible Universe. The explosion generated an intense burst of neutrinos—massless, chargeless particles emitted in nuclear reactions. On February 21, 1987, 100 billion neutrinos from the exploding star passed through the body of every person on the earth. Astronomers had theorized that most of the energy of an exploding star is carried off by neutrinos. The detection of the burst of neutrinos, exactly as predicted, gave the astronomers added confidence that their ideas regarding stellar births and deaths are correct.

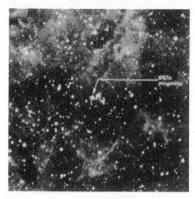 

1987 Supernova: Before   After

The Large Magellanic Cloud, site of a supernova in 1987 *(opposite)*.

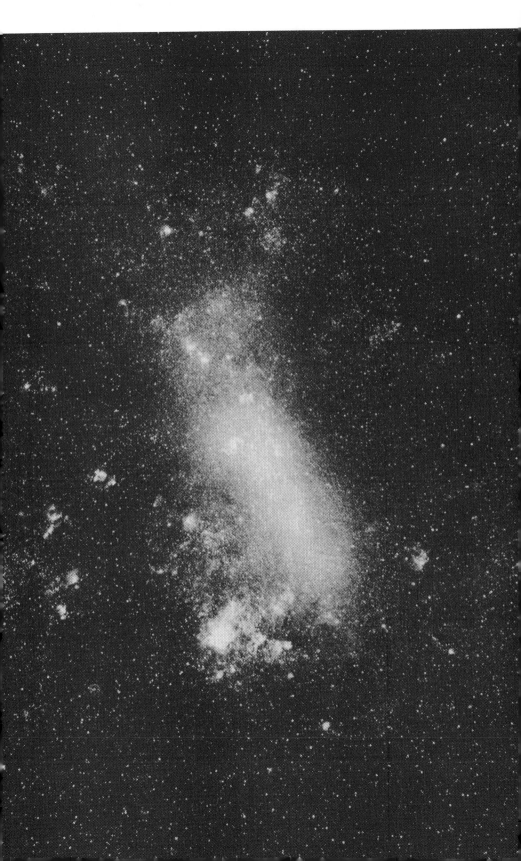

5

Black Holes and Quasars

Robert Oppenheimer was a famous American physicist who headed the atomic bomb project in Los Alamos during World War II. Before the war started, he and a student did a calculation the results of which amazed him. Oppenheimer used Einstein's theory of relativity to figure out what happens to a collapsing star at the end of its life. The result was "very odd," he wrote to a friend.

Oppenheimer had discovered that when a star comes to the end of its life and collapses, sometimes the collapse continues without stopping. The matter in the center of the star—all thousand trillion trillion tons of it—is squeezed from its original million-mile diameter down to a space the size of a house; then to the size of a golf ball; then to the size of a pin head; and then smaller, ever smaller.

Finally, the star's trillions on trillions of tons pile up at the center into a minute, almost unimaginably dense lump of matter.

The pull of gravity on the surface of this dense knot of matter is inconceivably great; in fact, gravity is so strong that nothing can escape from the squeezed star. Not even light can get away; like a ball thrown upward from the surface of the earth, and then pulled back to the ground by gravity, every ray of light from the star is pulled back into its interior. Since no rays of light can get out, the star is invisible. It has become a black hole in space.*

This is the meaning of the black hole. It is an enormously compressed object whose gravity is so powerful that nothing can escape. Everything inside the black hole is trapped there forever.

Any ray of light or material object that enters the black hole from the outside is also trapped; it can never get out again. The interior of the black hole is completely isolated from the outside world; it can swallow energy and matter, but it cannot send anything back. In effect, the material inside the black hole has been taken out of our Universe. It has become a universe of its own.

---

*According to Oppenheimer's calculation, in the first stage of the collapse the pressure in the center of the star squeezes negatively charged electrons together with positively charged nuclei to form neutrons. A giant ball of neutrons, called a neutron star, appears in the center of the star. Since neutrons take up much less space than separate electrons and protons, the neutron star is very compact and dense. In fact, a brick-sized chunk of matter from a neutron star would weigh 100 billion tons. If you put a brick of neutron star-stuff on the table, it would drill a clean hole through the table, the floor, the ground beneath, and straight through the center of the earth to the other side of the world.

That much was already known when Oppenheimer did his calculation. The unusual part of Oppenheimer's results was his finding that the collapsing star can go past the stage of the neutron star to form an object even stranger than a neutron star—stranger, in fact, than any astronomer had ever dreamed could exist in the heavens.

Suppose a spaceship passes by a black hole. As long as it does not come too close, it will be safe; the black hole's gravity will pull at it and bend its path somewhat, but nothing else will happen to it. However, if the spaceship heads directly for the black hole, it will enter it and vanish, never to be seen again.

What would happen to an astronaut who fell into a black hole? The properties of black holes seem to suggest that he would be crushed by gravity. In actual fact he would be torn apart, because the part of his body closest to the center of the black hole would be pulled by a gravitational force stronger than on any other part.

Suppose, for example, the astronaut entered feet first; then his feet would be pulled more strongly than his head, and feet and head would tend to separate. The astronaut would feel as though he were stretched on a rack; a few thousandths of a second after entering the black hole, he would be dismembered; after a few more thousandths of a second, the individual atoms of his body would be broken into their separate neutrons, protons, and electrons; finally the elementary particles themselves would be torn into fragments whose nature is not yet known to physicists.

Black holes are very strange objects. If they exist, they must surely be the most extraordinary objects in the Universe. The theory of relativity predicts that whenever a sufficiently massive star undergoes a supernova explosion, a black hole must be left behind, but theories have been wrong before. How can the prediction be tested?

A test would seem to be impossible, since black holes by their nature are invisible. However, recent discoveries made in space provide convincing evidence that black holes actually do exist. The discoveries were made by satellites equipped with X-ray telescopes, which collect and focus the X rays coming to the earth from outer space. These X rays from space do not reach the ground because the earth's atmosphere screens them out. That is why the telescopes that focus the X rays must be put into satellites.

These X-ray telescopes, circling the earth in orbit, revealed that many stars and galaxies in the heavens are also powerful sources of X rays. One intense shower of X rays seemed to be coming from a bright star. Yet careful observation showed that the X rays were not coming from the star itself, but from an invisible object very close to it.

There is reason to believe that the invisible object is a black hole circling around the bright, visible star. According to the astronomers' calculations, the gravitational pull of the black hole is so powerful that it tears streamers of gaseous matter off the surface of the bright star. As the gas approaches the black hole, it picks up speed; the atoms collide with one another, and the gas becomes quite hot. Near the surface of the black hole, the hot gas reaches a temperature of millions of degrees. At that temperature the gas radiates an intense stream of X rays. Apparently, it was these X rays that were detected by the satellite. That explains how a black hole can produce X rays that betray its presence—even though the black hole itself is invisible.

When a star collapses on itself at the end of its life, the black holes that are created by the collapse are rather small; they are only a mile or so in diameter. But the latest findings indicate that these are not the only kinds of black holes in the Universe. Astronomers have found evidence that the Universe also contains giant black holes, a billion miles in diameter, each containing the mass of a billion stars.

The evidence for giant-sized black holes comes from one of the most puzzling and remarkable discoveries ever made in astronomy. Some years ago, astronomers noticed a very unusual star in the heavens. At least, this point of light looked like a star. But a measurement of the distance to the "star" showed that it was two billion light-years from us—far beyond the boundary of our own galaxy. If this was an ordinary star, it would be very faint at that great distance—too faint to be seen. The fact that the star could be seen at all, in spite of its great distance, indicated that it must be enormously brighter than an ordinary star. When allowance was made for the star's great distance, its true brightness turned out to be equal to that of hundreds of billions of ordinary stars.

Soon, other examples of the strange stars were found. One was a trillion times brighter than an ordinary star. Clearly, these points of light were not stars.

The brilliant lights in the heavens that resembled stars but were far too bright to be stars became known as quasi-stellar objects, or *quasars* for short.

The extraordinary brightness of quasars was only one of their unusual properties. Even stranger was the fact that their enormous outpouring of energy seemed to be coming from a remarkably small region in space—smaller, in fact, than our solar system.

That was the real puzzle. How can an object as small as a solar system produce the energy of hundreds of billions of stars? If quasars produce the energy of hundreds of billions of stars, they must contain hundreds of billions of stars. But quasars occupy far too small a volume of space to contain that many stars.

Astronomers wondered if a force as yet unknown to science was generating the energy of the quasars—a force even stronger than the nuclear force that generates the energy radiated by an ordinary star. That would be a momentous finding, for up to now nuclear force has been the most powerful force known to man. The mystery deepened.

But then a clue to the nature of the strange quasars appeared. When the astronomers modified the instruments in their telescopes so that they could detect very dim objects, they saw the faint image of a galaxy around several of the quasars. At first astronomers had not noticed the galaxy surrounding each of these quasars, because each galaxy was so much fainter than the brilliant quasar at its center; if the astronomer set the time exposure in the telescope short enough to avoid overexposing the image of the bright quasar, the galaxy's image failed to appear in the photograph.

As soon as it was realized that quasars were located in the centers of galaxies, several astronomers thought of a solution to the mystery. What is extremely compact, yet can generate an enormous amount of energy? Only one object fills the bill. It seems almost too bizarre to be true, but—is it possible that an enormous black hole lurks in the center of each of these galaxies? Are quasars really giant black holes?

If quasars are giant black holes, the source of the quasar's enormous energy output becomes clear. The black hole sits in the center of its galaxy, surrounded by many stars. The stars circle around the black hole under the pull of its gravity. Gradually they spiral in toward it. As each star draws close to the black hole, its gaseous body is torn apart by the black hole's powerful gravitational force. The atoms of gaseous matter within the disintegrating star, picking up speed under the attraction of the black hole, move faster and faster. As the atoms approach the boundary of the black hole, they collide with one another. The collisions heat the gas, and the hot gas radiates energy into space. This energy is what we see when we observe a quasar.

The calculations show that if the black hole in the center of the galaxy is an ordinary-sized black hole—the kind that results from the explosion that marks the death

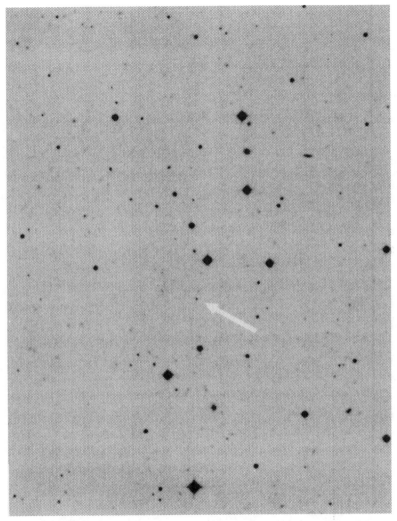

A DISTANT QUASAR. The small, black dot in the center of the negative photograph *above* is a distant quasar. The other black dots in the photograph are relatively nearby stars in our own galaxy. This quasar has the brilliance of 100 *trillion* stars, yet it seems faint compared to the many single stars nearby because it is 12 billion light-years away—far beyond the boundary of our galaxy and three quarters of the way to the edge of the visible Universe.

75

of a massive star—it will not produce enough energy to explain the brilliance of the quasars. But if the black hole is giant-sized—say, a billion times more massive than an "ordinary" black hole—it tears apart nearby stars with such vigor that it radiates huge amounts of energy into space—sufficient to account for the energy coming from even the most dazzling quasars.

So, the picture of a giant black hole lurking at the center of a galaxy explains the two remarkable properties of quasars. First, it accounts for the prodigious amount of energy coming from a quasar. Second, since black holes are exceedingly compact objects, it accounts for the fact that the energy comes from a very small region in space.

As soon as astronomers conceived the idea that giant black holes might exist in the centers of galaxies, they found it easy to understand, with hindsight, why the center of a galaxy was exactly where a giant black hole ought to be.

In the center of a galaxy, stars are much more closely packed together than they are anywhere else in the galaxy, because the galaxy's gravitational pull tends to draw them toward the center, and they pile up there. The density of the stars near the center of the galaxy can be millions of times greater than in its outer regions. Since the stars are packed together so closely in the center of the galaxy, they collide there frequently. In the outer part of a galaxy, collisions between stars are very rare; perhaps one occurs every billion years. But near the center, a collision between stars may occur every hour.

When two stars collide, they tend to fuse into one larger star in place of the two separate, smaller ones. Collisions being frequent in the center of the galaxy, this newly enlarged, double-sized star is likely to collide again in a short time with a third star, producing a still larger star as a result. Several collisions in a row lead to a very large and massive star—the kind that ends its life in an explosion, leaving a black hole behind.

That means the center of a galaxy should contain a considerable number of black holes. These black holes are not yet giant-sized; they are the "ordinary" black holes formed when a massive star explodes and dies. But Einstein's theory of relativity explains why a large number of "ordinary" black holes in the center of a galaxy are likely to amalgamate into a single giant-sized black hole.

The theory predicts that the diameter of a black hole is proportional to the amount of matter inside it. Thus, every time a black hole encounters another object and swallows it, the black hole grows bigger. Being bigger, the black hole is now even more likely to collide with and swallow other objects. A runaway process starts, in which the bigger the black hole is, the more likely it is to swallow other objects, and the more objects it swallows, the bigger it gets.

The runaway process continues until the biggest black holes have swallowed up all the smaller ones. Finally, the remaining black holes collide with and swallow one another. At the end, one giant black hole sits in the center of the galaxy.

After the giant black hole has formed in this way, the stars of the galaxy continue to circle around it, gradually drawing nearer. Once in a while, a star comes too close. Pulled in by the giant black hole's gravity, it is torn apart and consumed. The ravished star emits a great burst of energy in the final moments of its existence. These bursts of energy, coming one after the other as the stars close to the black hole are consumed, fuel the quasar's extraordinary energy output.

The story of quasars and giant black holes is nearly

77

complete. A quasar is a galaxy with a giant black hole at its center. The quasar's dazzling radiation is created by stars that feed, one by one, into the giant black hole. Each time the giant black hole tears a star apart, we see the quasar flare up as if another log has been thrown on the fire.

At first, the fire blazes brightly because the giant black hole has an ample supply of stars available for feeding. In other words, the brilliance of the quasar is very great in this early period.

But after a time, many stars in the inner part of the galaxy are gone because they have been torn apart and consumed by the giant black hole. After a relatively short interval of time, perhaps a few hundred million years, very few stars are left. With its source of energy gone, the quasar fades into darkness. Where the quasar once blazed, a galaxy of ordinary appearance remains—but with a quiescent black hole slumbering at its center.

This picture explains why quasars are so rare in the Universe. Earlier in the history of the Universe, quasars— that is, galaxies with active, star-eating black holes at their centers—may have been fairly common. In fact, many galaxies may have been quasars then, with giant black holes at their centers, consuming nearby stars. But in these galaxies the black holes have long since eaten all the stars within their reach, and are no longer active. The quasars have faded into ordinary galaxies.

In that case, why do we see any quasars at all? Why hasn't every quasar run out of fuel by now, and faded away? The answer is connected with the fact that most of the quasars we see are very far away. Because they are so distant, it takes a long time for the light from these quasars to reach us. Thus, we see these quasars not as they are today but as they were in the distant past, when the light from them had just started out on its journey to the earth.

But in that early epoch, the giant black holes at the centers of those galaxies had not yet consumed all the stars around them, and the quasars were still burning brightly.

Today those black holes are slumbering, and the quasars are dark. If we could see one of them as it is now, we would only see a galaxy—but we cannot see it as it is now, because it is so far away. We can only see it as it was a long time ago.

If this explanation is correct, many galaxies may have giant black holes lurking in their centers. If a galaxy is close to us, we see it as it is today; the giant black hole at its center is no longer active, and it looks like an ordinary galaxy. But if the galaxy is very distant, we see it as it was in the past, when its giant black hole was active—that is, when it was a quasar.

Astronomers have tried to invent other explanations besides black holes for the mysterious quasars. They feel the giant black hole is such an unbelievable object that its existence should be accepted only as a last resort. But nothing fits all the known facts as well as the black hole. And so one of the most remarkable chapters in the history of astronomy and science may be drawing to a close. Black holes—"the greatest crisis for physics of all time," as renowned physicist John Wheeler described them—must now be accepted as a reality.

ENERGETIC JETS OF MATTER.
Cygnus A *(right)* is a galaxy or
galaxylike object about 600 million light-years away that emits
an enormous amount of energy
in the form of radio waves. The
total amount of radio energy
emitted by Cygnus A is greater
than the energy emitted in the
form of light by all the stars in
our galaxy.

Radio telescopes can collect and focus the radio waves from
Cygnus A to form a radio image, just as ordinary telescopes collect
and focus light to form a visible image of the galaxy. The radio
image of Cygnus A *(below)* reveals a remarkable phenomenon:
Cygnus A—the small, white dot at the center of the radio image—
lies midway between two huge jets of energetic matter, which
appear to squirt out of opposite sides of the galaxy. Each jet is
approximately 150,000 light-years in length. The source of energy
for these jets appears to be a giant black hole in the center of
Cygnus A. Jets like this have been observed emerging from more
than 100 galaxies and quasars.

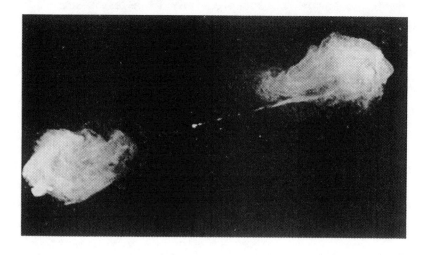

EVIDENCE FOR GIANT BLACK HOLES IN NEARBY GALAXIES. The Andromeda Galaxy, or M31 *(above)*, is only two million light-years, or 12 million trillion miles, away and is one of our closest galactic neighbors. Measurements of the speeds of stars in the Andromeda Galaxy reveal that they are moving very rapidly. In fact, these stars would leave the galaxy and fly into space if they were not held there by the gravitational pull of an invisible object at the center.

From the speeds of the stars it can be calculated that the mysterious object is tiny in its dimensions, but massive—roughly 50 million times as massive as the sun. The massive but extremely compact object in the center of Andromeda appears to be a black hole.

III

Exploring Our Solar System

6

The Sun's Family

The time is four and a half billion years ago. The Universe is ten billion years old. Several billion stars in our galaxy have already lived, died, and exploded, and the debris of their bodies has mixed with the primordial gases of the Cosmos. A substantial supply of elements like carbon, oxygen, silicon, and iron—made in stars, and necessary for making planets and living things—has accumulated in the Universe. The stage is set for the appearance of the sun and its family.

Now a cloud gathers out of the swirling mists of the galaxy. The atoms in the cloud will later form the bodies of the sun, the earth, and the creatures that walk on the surface of the earth. It is the parent cloud of us all.

At first, the parent cloud is very large and tenuous, extending over many trillions of miles. With the passage of time, gravity draws the atoms of the cloud together, and the cloud contracts. As it contracts, it becomes denser and hotter. After some 10 million years, the center of the cloud flares up in a chain of nuclear reactions. This moment marks the birth of the sun.

While the sun is forming in the hot center of the cloud, smaller knots of condensed matter appear in the cloud's outer and cooler regions. These clumps of matter are also held together by their own gravitational attraction. Later, according to the modern theory of the origin of the solar system, the smaller knots of condensed matter become the planets.

This picture provides a good explanation for the formation of the largest planets in the solar system—Jupiter, Saturn, Uranus, and Neptune. These are the so-called giant planets, which are made mainly of hydrogen and helium— the same gases that make up the mass of the sun and most stars. There is no question that the giant planets formed in the same way as a star. In fact, if Jupiter, the largest of the giant planets, were larger still, it could be a small star, burning nuclear fuel, and shining brightly by the nuclear energy produced in its center.

However, the picture of a planet forming like a star does not work as well for the earth or its sister planets— Mercury, Venus, and Mars. These bodies, known as the earthlike planets, are quite small compared to the giant planets; the earth, for example, is only one three-hundredth as massive as Jupiter. It is the small mass of the earth that creates a problem. The earth is just not massive enough to have been held together initially by the force of its own gravity. It could not have condensed out of the gases of the parent cloud as Jupiter and the other giant planets did.

Then how *did* the earth form? That is one of the minor mysteries in the scientist's story of Genesis. Astronomers believe that something like the following occurred. In the beginning, there was a cloud of gaseous matter, with the young sun at its center. Gradually, as the years went by, this cloud lost its heat to space. When it had cooled sufficiently, the atoms within the cloud began to stick together to form tiny grains of solid matter.

The first grains of matter to appear in abundance were bits of iron. They appeared first because iron atoms are more strongly attracted to one another than most atoms, and are more likely to stick together in small clumps. After iron, the next bits of solid matter to appear were grains of rocklike material. Now the bits of rock and iron circled the sun, immersed in the lighter gases of the parent cloud. Some collided and stuck together, or were drawn together by gravity, forming larger bodies up to perhaps a mile in diameter.

Then the sun flared up in a violent outburst, as young stars are prone to do. The flare-up blasted the light gases out of the inner part of the solar system. But the bodies of rock and iron, each containing trillions of individual atoms, were too heavy to be blasted away by the streams of particles and radiation from the sun. They continued to circle around the sun after the gases were gone—each a miniature planet in its own right. Later, those balls of rock and iron collected together to form the earth and its sister planets—Mercury, Venus, and Mars.

This account of the creation of the earth is generally accepted, but one interesting question remains. What forced the chunks of rock and iron together to form a full-sized planet? Why did they not continue to circle the sun as separate pieces of earthlike material?

The answer is not clear, but as far as astronomers can tell, something like the following must have occurred. Now and then, as the of pieces of rock and iron moved around the sun, neighboring bits of matter came close to one another by accident. Sometimes, when this happened, gravity drew the pieces still closer together. As they came closer, the gravitational attraction grew stronger, and drew them closer still. A runaway effect devoloped, in which, under the force of gravity, a knot of condensed matter, formed by accident, grew into the nucleus of a planet.

This process went on over a period of many million years, proceeding very slowly at first, and then with rapidly increasing momentum in the final stages. At the end, nearly all the bits of matter in the inner part of the solar system were gathered into the earthlike planets, and only a few atoms of gas remained in the space between.

The result was a solar system with two kinds of planets. The earthlike planets, relatively small bodies made of rock and iron, occupied the inner region of the solar system, closest to the sun. The giant planets, massive objects made mainly of hydrogen and helium, occupied the outer region.

Between the two groups of planets lay a broad, planetary no-man's-land, 100 million miles wide. In this no-man's-land, where another planet ought to be, there were only fragments of planetary matter—bits of rock and iron of various sizes—circling the sun in a diffuse ring. These fragments of a planet are called *asteroids*.

The asteroids move around the sun in a broad zone called the asteroid belt. Many hundreds of thousands of asteroids probably exist in the asteroid belt, although less than 10,000 have been found thus far by astronomers. The largest is Ceres, whose diameter is nearly 500 miles. Three other asteroids—Pallas, Vesta, and Hygeia—have diameters greater than 200 miles. But most asteroids are far smaller, ranging down to the size of sand grains.

Why does the asteroid belt contain only pieces of planetary material, instead of a full-sized planet? At one

FORMATION OF THE PLANETS *(above)*. In the first stage, a cloud formed out of the gas and dust of space, drawn together by gravity. The hot gas at the center of the cloud gave rise to the sun. Farther out in the cloud, the temperature was lower and grains of solid matter condensed.

The tiny grains circled around the newborn sun. Now and then they collided and stuck together like snowflakes, or were drawn together by gravity, gradually accumulating into larger bodies. These were the nuclei of the earthlike planets and their moons. In the final stage, the large bodies swept up the remaining bits of matter in the cloud, and grew into full-sized planets and moons.

time it was thought that a planet might have formed at that distance from the sun and then exploded; or perhaps a collision had occurred between two planets, and the asteroids were the relics of the catastrophe. According to modern ideas, neither explanation is correct. Jupiter is the culprit.

When the solar system was young, before the earthlike planets had formed, small bits of iron and rock circled the sun everywhere, held captive by its gravity. In the inner part of the solar system, these bits of iron and rocky material collected to form the earth and its sister planets. But, farther out, beyond the orbit of Mars, the process of making a planet was disrupted by Jupiter. There, in what is today the asteroid belt, fragments of rock and iron started to collect into larger pieces and then into a full-sized planet. But before that happened, Jupiter's gravitational pull disturbed their paths. Because Jupiter is so massive, the force of its gravity is greater than that of any other body in the solar system except the sun itself. As a result, the asteroids, pushed and pulled by Jupiter's gravity, crashed into one another violently as they traveled around the sun, instead of colliding gently and coming together under the force of gravity. That is why the pieces of matter in the asteroid belt never accumulated into one large planetary body.

Jupiter's gravity is still pulling asteroids out of their orbits today. Occasionally it sets one asteroid on a collision course with another. When a collision occurs, the shattered fragments of the two asteroids leave the scene, traveling in many different directions and with different speeds. Some fragments are hurled farther out into the solar system. Other asteroid fragments are knocked in toward the sun by the collision. A few of these land in orbits that cross the orbit of the earth.

Sometimes, one of the asteroids that crosses our path actually collides with the earth. As this asteroid passes through the earth's atmosphere, the friction of its passage through the air heats it to incandescence and it leaves behind it a long trail of vaporized matter. At night, the glowing trail can be seen clearly as a bright streak against the blackness of the sky. These glowing trails are called shooting stars or falling stars.

If the asteroid that hits the earth is small—the size of a grain of sand or a pebble—it burns up in the atmosphere without reaching the ground. But some asteroids are big enough to survive the passage through the atmosphere and reach the ground more or less intact. An asteroid that hits the ground is called a *meteorite,* and may wind up in a museum. The passage of a large meteorite through the atmosphere creates an awesome fireball that makes a spectacular sight, visible in the daytime.

When very large meteorites strike the ground, they leave a crater as a scar of the impact. The Arizona meteorite crater is the result of a collision with a 300,000-ton meteorite—a visitor from the asteroid belt—that hit the earth about 30,000 years ago.

Every few million years an extremely large asteroid collides with the earth. This has not happened in historic times, so we have no experience of the consequences of that event. However, it must be a major catastrophe for life on the planet since a piece of rock a mile in diameter delivers the explosive energy of a million hydrogen bombs when it hits the ground. Some astronomers and geologists believe that the dinosaurs became extinct 65 million years ago because a large asteroid hit the earth at that time.

We are confident that such cataclysmic collisions have taken place, because we have a good estimate of the number of asteroids that can collide with the earth. About

1,000 asteroids that are more than a mile in diameter cut across the earth's orbit, and about 20 that are more than ten miles in diameter. From these numbers it is easy to calculate the probability of a collision with the earth. The calculation reveals that one-mile pieces of rock hit the earth every million years or so, and ten-mile pieces hit every 100 million years.

In 1937 Hermes, an asteroid roughly 10 miles in diameter, passed by the earth at a distance of 400,000 miles. On the scale of distance in the solar system, this was a very close miss. Hermes was tracked for five days when it went by in 1937, and then it was lost. We have not seen it since. It may strike the earth on another pass, not too far in the future.

On March 23, 1989, the inhabitants of the earth again had a close call when another large asteroid went by the planet at a distance of less than 500,000 miles. No one had suspected the existence of this asteroid. If it had collided with the earth, it would have released an amount of energy equivalent to the explosion of 40,000 hydrogen bombs.

The asteroid came out of the heavens with the sun behind it, and could only be seen after it had already passed the earth and was receding into the night sky. Astronomers tracked it for some days and then, like Hermes, it was lost. Either the new asteroid or Hermes may strike the earth on another pass, not too far in the future. The impact is most likely to be in the oceans, which cover three quarters of the area of the globe. If that happens, huge tidal waves may kill millions of people living in coastal regions.

•

Now the story of the solar system and its origin is nearly complete. The inner part of the solar system is occupied by the earthlike planets—Mercury, Venus, Earth, and Mars—

formed from bits of rock and iron. Beyond the earthlike planets lie the giant planets—Jupiter, Saturn, Uranus, and Neptune—huge spheres of partly gaseous matter resembling small stars in their compositions. Between the two groups of planets, innumerable fragments of matter circle the sun in the asteroid belt.

But one planet—neither an earthlike planet nor a giant planet—still remains to be described. Beyond Neptune, outermost of the giant planets, lies the frozen world of Pluto, the ninth and outermost planet in the solar system. Pluto is a small planet, made of roughly equal amounts of ice and rock, and unlike either the earth or the giant planets. It circles the sun at a distance of three to four billion miles, and takes 248 years to complete one turn around our star.

Pluto probably formed in the same way as the earthlike planets, out of small bits of solid matter that collected around the sun in the early years of the solar system. But because of Pluto's great distance from the sun, these bits of matter included bits of ice, in addition to pieces of rock and iron. In fact, it was so cold at Pluto's distance from the sun that not only did water freeze into ice, but methane also froze. Methane is a familiar substance on the earth, normally present in the form of a gas, and sometimes known as cooking gas. Methane does not become a solid until the temperature drops to 250 degrees below zero Fahrenheit. Pluto has an abundance of methane; caps of methane ice cover the poles, and the planet's atmosphere also consists mainly of methane.

Although Pluto is very small for a planet—considerably smaller, in fact, than our moon—it was discovered about 10 years ago that Pluto has a moon of its own, even smaller, called Charon after the boatman who ferries unfortunates across the River Styx to the underworld.

The discovery of Pluto's moon, Charon, led to a great

93

puzzle—the mystery of the tenth planet. That story begins in the nineteenth century, when an astronomer and a mathematician noticed that the giant planet Uranus, which lies beyond Jupiter and Saturn, was moving in a somewhat peculiar fashion as it orbited the sun, as if it were being tugged at by an unseen body. They analyzed the peculiar motions and decided they were caused by a previously undiscovered planet.

At that time, the world knew of only seven planets; Neptune and Pluto were still to be discovered. Astronomers calculated where the unknown planet should be, and sure enough, a year later they found a planet close to the spot they had predicted. They named the new planet Neptune.

As soon as Neptune was discovered, astronomers calculated the orbit of Uranus again, this time allowing for Neptune's influence. They expected to find very good agreement between their calculations and the measured orbit of Uranus. But several decades and untold pages of algebra and arithmetic later, they found, to their surprise, that the agreement was not very good; they still could not account for all the peculiarities in the orbit of Uranus. Neptune explained some of these peculiarities, but it did not provide the entire explanation. Some other force, in addition to Neptune, was tugging at Uranus.

That finding set off a search for still another unknown planet. This time the search lasted 25 years. It ended in 1930, when a young American astronomer found Pluto, the ninth and last planet to be admitted to the sun's family.

This would be the end of the story of the planets, except for still another strange development that grew out of the discovery of Charon, Pluto's moon. Charon enabled astronomers to calculate the mass of Pluto, which they had only been able to guess at before then. When a planet has a moon, the planet's mass can be calculated from the speed at which the moon circles the planet in orbit. It is easy to see why that is so. If the planet is very massive, it exerts a strong gravitational pull on its moon, and the moon must go around in its orbit at a high speed to avoid falling into the planet. On the other hand, if the planet is small, its gravitational pull is weak, and that means its moon revolves around it slowly; for if the moon moved too quickly, it would break loose from the planet's weak gravitational pull and fly off into space.

Using this reasoning, the astronomers calculated Pluto's mass from the speed with which its moon revolved around it. They found, to their surprise, that Pluto is a tiny planet, only a few thousandths of the mass of the earth.

But a planet as tiny as that cannot exert enough force on Uranus to explain its peculiar motions. To pull at Uranus strongly enough to account for its motion, Pluto would have to be about 100 times more massive than it actually is.

So the search is on again for another planet—Planet X, which will finally account for the peculiar motion of Uranus. If Planet X exists, it must be at least twice as massive as the earth, located in an orbit that could be as far out as 10 billion miles from the sun, and take nearly a thousand years to go around the sun. This mysterious body—the tenth planet—awaits discovery.

But even the tenth planet—if it is found—will not mark the outer boundary of the solar system. Much farther out lie the millions of small, icy bodies called *comets*.

The comets also circle the sun, and belong to the sun's family. Their distances from the sun, however, are generally very great, as much as a trillion miles, and it may take a comet a million years to complete one turn around the solar system.

The existence of the comets is another minor mystery of the solar system. It is easy to understand why comets should be made mainly of ice, because they formed in the outermost and coldest reaches of the parent cloud, farthest from the warming rays of the sun. But why they formed at all is a puzzle.

The most spectacular feature of a comet is the long tail that forms behind it as it sweeps in toward the sun. (The name is derived from the Latin word *cometa*, which means "long-haired.") The fact that comets are made mainly of ice explains why they grow these tails when they approach the sun. As the comet's orbit carries it past Jupiter and into the inner part of the solar system, the sun's rays warm its surface and vaporize some of the frozen material. The vapor, carrying with it small grains of rock and dust that were imbedded in the comet's ice body, streams out behind the comet in a trail that may be millions of miles long. The trail glows brightly against the night sky, making the close passage of a comet one of the most striking sights the heavens afford.

Some tiny grains of rocky matter and dust from the comet's tail are left behind as it passes around the sun. If the earth's orbit carries it through this swarm of comet-tail particles, many particles burn up in the earth's atmosphere. Each one leaves a glowing trail—a falling star. Together, the trails make a magnificent display, known as a *meteorite shower*. The particles in the tail of Halley's Comet produce a meteorite shower that is seen every year around the night of October 21.

Some comets have been dislodged from the long, sweeping orbits that take a million years for one pass around the sun, and into closer orbits in which they go around the sun in a matter of years or decades. Here, again, the culprit is Jupiter, whose gravity can disturb a comet's orbit if the intruder comes too close. The most famous of these "short-period" comets is Halley's Comet, which visited the earth in 1910 and came back in 1986. The latest visit gave astronomers their first chance to take a close look at a comet's "nucleus"—the frozen body that provides the materials for the comet's tail—from a passing spacecraft.

Astronomers and space scientists were avidly curious to see what the nucleus of a comet actually looked like. They found that the nucleus of Halley's Comet is about five by ten miles in size, with a lumpy shape variously described as that of an avocado, a potato, or a peanut. Its density is considerably less than the density of water, indicating that it is a porous and fluffy body, apparently made of ice and such substances as frozen carbon dioxide, ammonia, and methane. There are also bits of rocky material. The surface of Halley's Comet turned out to be jet black— blacker than coal. It reflects less than four percent of the light incident on it, compared to seven percent for anthracite coal, and is the blackest object known in the solar system.

The blackness results from a layer of carbon-rich material, similar to soot, that covers the Neapolitan sherbet of ice and rock making up the body of the comet. The layer of soot also contains organic molecules—the building blocks of life. However, conditions in the outer regions of the solar system, where comets form and spend most of their lives, are such that a comet is very unlikely to contain living organisms.

Planets, asteroids, and comets—the creation of the solar system is nearly at an end. Only the moons of the planets remain to be fitted into the picture.

Why do planets have moons? The answer is clear for the large moons in the solar system, such as Jupiter's four largest moons. Jupiter possesses these moons for the same reason the sun possesses planets. Jupiter's large moons are nearly as big as planets, and probably formed around Jupiter in the same way the earthlike planets formed around the sun, growing out of bits of planetary matter circling around Jupiter. Jupiter, in fact, resembles a miniature solar system—except that Jupiter, the massive body at the center of this "solar system," is not a true star like the sun because it does not have a nuclear fire burning at its center.

The small moons in the solar system have a different origin. They include the two moons of Mars, Phobos and Deimos—potato-shaped chunks of rock, roughly ten miles across—as well as dozens of small moons that revolve around Jupiter and the other giant planets.

All these small moons are probably captured asteroids, snared from the asteroid belt. When a fragment of rock in the asteroid belt is perturbed by the pull of Jupiter's gravity and deflected out of its normal orbit, it may pass by a neighboring planet, such as Mars, at just the right distance—neither so close that it crashes into the planet's surface, nor so far away that it whips past it and flies off into space.

In that case, the fragment can settle nicely into the orbit of a satellite, captured by the planet's gravity. The capture of asteroids is a likely explanation for the moons of Mars and the dozens of smaller moons that circle the giant planets.

This leaves only the earth's moon. It fits neither explanation. Our moon cannot be a planet in the earth's "solar system," because the earth is too small to have a family of planets. And our moon is made of such different materials from the asteroids as to indicate that it could not have been captured from the asteroid belt. In fact, no good explanation of the moon's origin has been provided. On a cosmic scale, the mystery of our moon's origin is minor, and yet it is provoking. Harold Urey, father of lunar science, studied the problem and gave up, saying, "It is easier to pretend the moon is not in the sky than to explain how it came to be there."

THE ORIGIN OF THE MOON. Harold Urey discusses, and discards, three theories on the origin of the moon at a NASA conference. The theories listed on the blackboard are *capture* (the moon is drawn into an orbit around the earth during a chance encounter), *binary formation* (the moon and earth form at the same time as a double planet), and *fission* (the moon is originally part of the earth and separates from it by a kind of twinning process.

JUPITER: A MINIATURE SOLAR SYSTEM. Jupiter, largest of the nine planets, has a family of planetary bodies circling around it, including four large, planet-sized moons and numerous smaller moons. Jupiter is seen here in the *background*, with three of its four largest moons—Io, Europa, and Ganymede—in this montage of photographs obtained by the NASA spacecraft *Voyager*.

Io *(upper left)* is composed almost entirely of rock and iron, with little or no water. *Voyager* has observed volcanoes erupting on Io. Io and the earth are the only volcanically active bodies known in the solar system. Ganymede *(lower left)* is made of ice and rock in roughly equal proportions. This moon is larger than the planet Mercury. Europa *(center)* is entirely covered with ice and has an extraordinarily smooth surface, its tallest mountains being only a few hundred feet high.

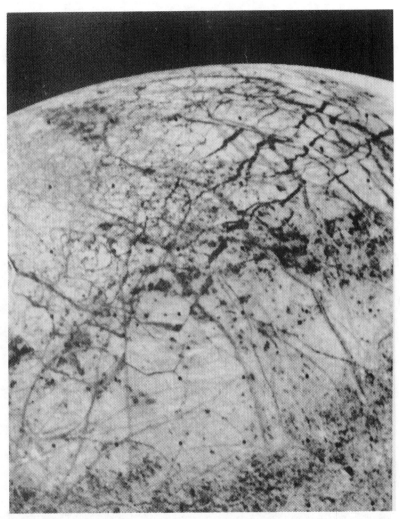

EUROPA. This *Voyager* photograph of Europa *(above)*, one of Jupiter's planet-sized moons, at close range reveals myriad jagged lines colored various shades of orange and brown that run across the surface in every direction. Their origin is a mystery. However, they have the appearance of cracks in the global ice sheet that were filled by muddy water, welling up from below and freezing at the surface. One school of thought holds that an ocean of water lies under Europa's ice-covered crust, possibly harboring a form of life.

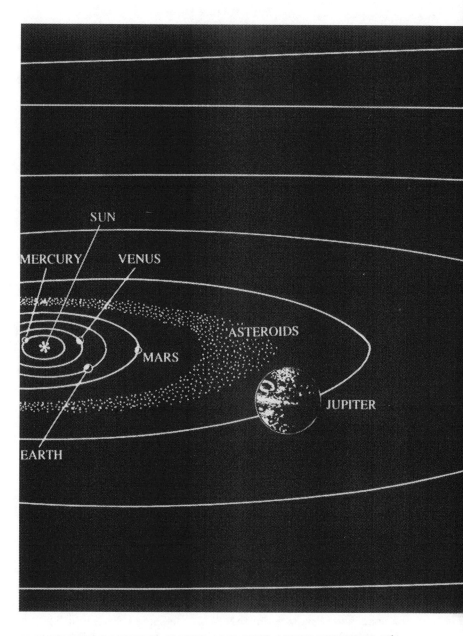

THE ORBITS OF THE PLANETS AND THE ASTEROID BELT. The orbits of the planets are shown to scale. A gap appears in the even progression of the orbits between Mars and Jupiter, suggesting that another planet should exist at this distance from the sun. However, only fragments of planetary matter—the asteroids—are found there. These fragments never collected to form a planet because Jupiter's

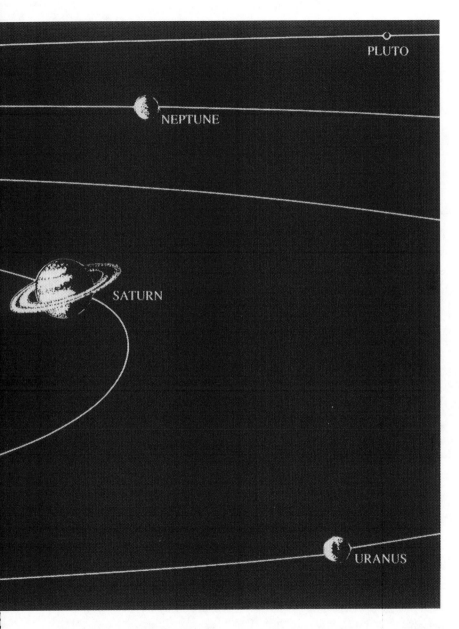

gravitational pull disturbed their motions, causing them to collide with one another repeatedly, and disrupting the process of accumulation into a single large planetary body.

Some asteroids have been pushed or pulled by Jupiter into orbits that cut across the orbit of the earth. Now and then, a large one hits the earth, releasing the energy of a million hydrogen bombs.

THE ARIZONA METEORITE CRATER. This crater, nearly a mile across and 600 feet deep, was formed approximately 30,000 years ago when a large piece of iron and rock from the asteroid belt, weighing a million tons, struck the earth at a speed of 20,000 miles an hour. The collision released the energy of a dozen hydrogen bombs.

Asteroids of this size hit the earth every few hundred thousand years. Larger asteroids, some nearly as large as the island of Manhattan, strike the earth every hundred million years or so. If a large asteroid strikes a densely populated region, such as one of the major cities, millions of lives will be lost. The chance that an asteroid will hit a large city is roughly one in 10,000—small, but not zero.

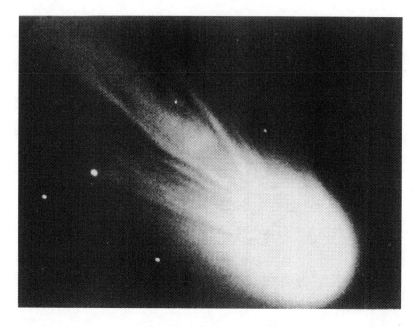

COMETS. The near approach of a comet is one of the most extraordinary sights in the sky. The comet shown here is Halley's Comet, which last visited the earth's neighborhood in 1986.

The streaming gases that form a comet's impressive visual display come from a small, solid body called the nucleus—a ball of ice, rock, and frozen gases—whose orbit periodically carries it from the outer region of the solar system in toward the sun. As the nucleus of the comet nears the sun, the frozen gases are vaporized and stream out behind it in a spectacular, glowing tail. When Halley's Comet is approaching the sun, it loses many tons each second in this way, but can last for many thousands of years, since its mass is nearly 200 billion tons.

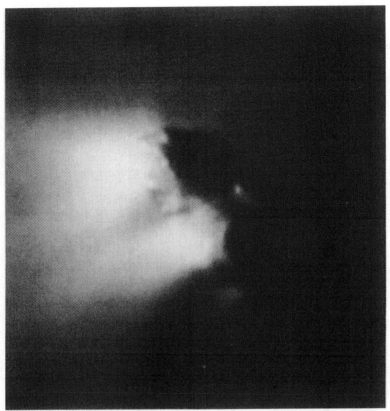

THE NUCLEUS OF HALLEY'S COMET. The nucleus of a comet had never been seen prior to 1986, when European and Soviet spacecraft passed close by the nucleus of Halley's Comet. The European spacecraft *Giotto* took a striking photograph of the comet's nucleus. It turned out to be a potatolike structure some 10 miles long, covered with a baked-out, jet black crust rich in carbon compounds.

The nucleus rotates every seven days and wobbles like a top as it rotates. The photograph shows two clouds of vapor streaming to the left from large holes in the crust of the nucleus. Whenever the spinning, wobbling motion of the nucleus turns these holes toward the sun, its rays penetrate into the body of Halley, vaporizing the ice, which explodes into the geysers of gas and dust visible in the photograph.

7

Many Earths Circling Many Stars

When I was a young student, I was taught that the earth and its sister planets were formed long ago in a collision between the sun and a passing star. The force of the collision tore huge streamers of flaming gas out of the bodies of the two stars. As the intruding star receded into the distance, some streamers of gaseous material were captured into circular orbits around the sun.

One of those long streamers of hot gas condensed into a molten mass, on whose surface a crust formed. Gradually, the mass cooled and hardened with the passage of time. This molten, slowly cooling mass was the earth. Other masses cooled and hardened to form the remaining planets of the solar system.

It is easy to calculate how often this could have happened. The likelihood of a collision between two stars depends on how big the stars are and on the distance between them. It turns out that the size of a typical star is a million miles, but the average distance between stars is enormous—some trillion miles. In other words, stars are far smaller than the distance that separates them from their neighbors.

That means the space through which the stars move is nearly empty, and collisions between stars are extremely rare. In fact, the calculations show that only a few such collisions have occurred in the entire history of our galaxy.

Therefore—if this is the way solar systems are formed—only a few solar systems exist in our galaxy.

Since solar systems and planets seem to be necessary for life to start, it follows that ours may be the only life in the galaxy. In fact, mankind, the intelligent creatures, may be alone in the galaxy, and perhaps alone in the Universe.

An entirely different prediction comes out of modern ideas on the birth of the stars. According to these ideas, whenever a star begins to condense out of the gases of space, planets are likely to condense around it. If this theory is correct, the earth is duplicated again and again in the Universe, like a design on wallpaper.

That conclusion has important implications for the existence of life elsewhere in the Universe. Even if the chance of life developing on another planet is small—as small, say, as one in a million—the number of stars and planets is so great that there must be an enormous multitude of inhabited solar systems around us. The Universe must be teeming with life of all shapes, sizes, and levels of intelligence. All speculations about extraterrestrial life rest on the assumption that this is so.

Unfortunately, no one has been able to confirm that interesting prediction directly. That is, no one has ever looked at a distant star through a telescope and seen, close to the star, a faint point of light that could be a planet. The reason is that the earth's atmosphere blurs the images of stars seen in the telescope, so that instead of being sharp points of light they are diffuse, luminous circles. Since the planet is quite close to the star, this luminous circle tends to spread over the planet's image, and prevents it from being seen.

The difficulty caused by the earth's atmosphere can be avoided if the telescope is sent up into orbit in a satellite, because then the telescope is far above the atmosphere and its images are very sharp. A large telescope in space can bring the light from a star to a sharp enough focus so that the faint image of a nearby planet may be revealed—if the planet is there. Telescopes in space have a better chance of discovering planets around other stars than telescopes on the ground.

The search for planets in other solar systems will be one of the most important tasks assigned to the Hubble Space Telescope. The Space Telescope will carry out the search with the help of an ingenious stratagem by which the edge of the moon is used as a mask to block the glare of the light from the central star. This is possible because the images produced by the Hubble Space Telescope are so sharp. If its images were as blurred as the images formed in telescopes on the ground, the light from the central star would spill over the edge of the screen and obscure the very faint light from the planet.

Although planets circling other stars cannot be seen readily through telescopes located on the ground, their presence can be detected from the ground by an indirect method. As a planet goes around a star, the planet's gravity pulls at the star and causes a little wobble in its path as it moves through space. The wobble is not very great because the star is much more massive than the planet; therefore, it does not move much in response to the planet's pull. The situation is like that of a large, heavy dance partner—the star—waltzing a small partner—the planet— across the dance floor. The large, heavy partner moves very little, and the small partner moves a great deal.

But the star's wobble, although slight, can be detected, because it has an effect on the wavelength of the light emitted by the star. When the planet is on the side of the star nearer to us, it pulls the star in our direction. This causes the light waves coming toward the earth to be compressed, which means their wavelength is decreased. On the other side of the planet's orbit, it pulls the star away from us, and the light waves coming to us from the star are stretched. The result is a rhythmic change in the wavelength of the light from the star, alternating from short to long and back to short again, every time the planet goes around the star in its orbit.

In 1987 astronomers reported that they had found just this rhythmic shift in wavelength in the light from several nearby stars. Apparently, these stars are being tugged at by unseen planets. The planets doing the tugging appear to be the size of Jupiter or larger. Smaller planets the size of the earth may also be present, but these indirect measurements are not sensitive enough to reveal them. However, it seems likely that if other solar systems contain large, Jupiter-sized planets, they also contain small planets like the earth, just as our solar system does.

Still more evidence for other solar systems came to light recently. A NASA satellite made the remarkable discovery that several stars near us are surrounded by objects that appear to be planets in the process of being born, but not yet fully formed. This NASA satellite was designed to search the heavens for objects producing infrared radiation, or heat waves. It recorded a faint emanation of heat from the vicinity of the star Vega, one of the sun's closest neighbors. The heat seemed to be coming from a ring of dust and small particles circling Vega in orbit. The particles appeared to range from the size of a grain of sand to the size of a brick.

According to the latest ideas on the origin of planets, a ring of particles of this kind, circling a star in orbit, is believed to be the first step in the creation of a planet like the earth. The Vega discovery is the first evidence that these ideas are correct; earthlike planets really do form by the coalescence of bits of matter of various sizes.

Following up on this discovery, astronomers have found that no less than a third of the stars near us emit the telltale heat radiation that indicates the birth of a family of planets. That is a high enough percentage to make it appear that stars with planets are indeed commonplace in the Universe, as modern theories on the birth of stars suggest.

A few short years ago, the idea that planets form as a natural accompaniment to the birth of stars, and are common in the heavens, was only a theory. Now it is close to an established fact. Mankind must become accustomed to the notion that the earth is not unusual, but is one among a vast multitude of similar planets. The new evidence indicates that billions of earths exist in our galaxy, and many billions more in the observable Universe.

Four centuries have passed since Giordano Bruno suggested that the sun is one of an infinite number of stars, and planets like the earth are commonplace. Burned at the stake in 1600, he was made to suffer for his originality. Only now, for the first time, has science obtained evidence that Bruno was right. Planetary worlds really do exist on other stars; they are probably made of the same materials as the earth; and they appear to be as numerous as the grains of sand in all the oceans of the earth.

BIRTH OF PLANETS. This photograph of the star Beta Pictoris, 53 light-years from us, shows a disc of particles circling the star. The disc is edge on to our line of sight, and extends out 40 billion miles from Beta Pictoris. The star itself is hidden by a circular mask at the center, which blocks the glare of light from Beta Pictoris, enabling the relatively faint light from the particles to be seen.

The photograph confirms the idea that a disc of particles is the first stage in the formation of planets. The density of the particles in the ring around Beta Pictoris suggests that planets may have already started to form there.

114

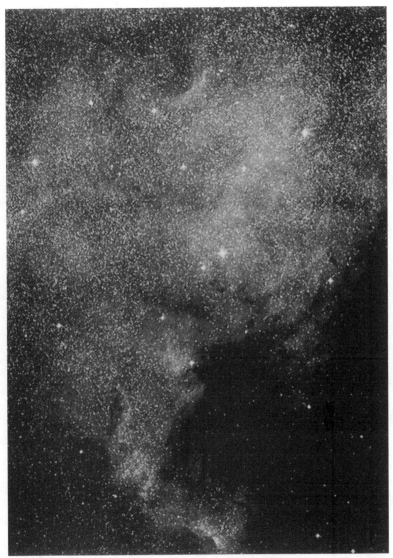

A MULTITUDE OF SUNS. A photograph of one small region in our galaxy, called the North American Nebula, conveys a sense of the enormous multitude of stars and planets in the visible Universe. Every point of light in the photograph is a sun; according to the latest evidence, many of these suns have families of planets.

8

The Search for Life

At some point in the first billion years of the earth's existence, life appeared in the waters on its surface. How that life came to the earth is one of the supreme cosmic mysteries.

We know the earth was lifeless at the start. Debris of all sizes, bombarding the planet during its birth process, had heated and melted its outer layers. An ocean of lava covered the surface. If any life existed in the bits of matter out of which the earth formed, that life surely was destroyed in the firebath of the earth's creation.

Slowly the molten surface of the earth cooled and solidified. Gases trapped in the earth's interior when it formed bubbled up to the surface to create the planet's first atmosphere. Steam, released from the lava, cooled and condensed into pools of warm water on the surface.

Somehow, life emerged from those pools of water in the earth's first billion years. Scientists are not sure they

117

know how that happened. They can only offer a guess, based on the fact that all life on the earth—animals, plants, and microbes—is made out of the same basic molecules. These molecules, called amino acids and nucleotides, are the building blocks of living matter. Like pieces of an erector set that can be used to build many different things, the molecular building blocks of life can be put together in different combinations to make a tree, a mouse, or a man.

Laboratory experiments show that the building blocks of life were probably formed in great numbers in the atmosphere of the earth when it was a young planet. Draining out of the atmosphere into the oceans, the basic molecules created a nutritious broth of life-giving materials. Collisions occurred between neighboring molecules in the broth, and sometimes two or three small molecules stuck together to form a larger one. Then another molecule collided with that one and stuck, and then perhaps still another. In this way, very large molecules were formed by a succession of random collisions.

Millions of years went by; countless collisions took place; gradually, many sizes and shapes of large molecules were built up in random encounters. Eventually, in this great variety of molecules formed by chance collisions, a kind of molecule appeared that had the magical ability to divide and make copies of itself.

That was the first self-reproducing molecule. It was the parent; the copies were its daughters. The daughter molecules inherited the parent molecule's magical trait; they divided, and made copies of themselves also. One molecule became two; two became four; four became eight; eight became 16; and then 32; and then 64. Soon, the offspring of the self-reproducing molecule were more numerous than any other kind of molecule in the waters of the earth.

The appearance of a molecule that could divide and reproduce itself was the critical step from nonlife to life. Without a molecule like this inside a cell, the cell cannot divide; without cell division, an organism cannot grow, and cannot produce offspring.

The self-reproducing molecule was the beginning of that succession of countless generations that led, by imperceptibly small changes in each generation, from the simple to the complex—from microbes to man. It was the start of parenthood; it was the start of biological evolution; it was the beginning of life. Today, the descendants of these self-reproducing molecules are everywhere on the earth; they are the molecules called DNA, which lie in the center of every living cell.

What scientific evidence supports this remarkable theory of the origin of life? There is very little. Laboratory experiments have given us clues to how it all started; they have created the building blocks of life; but they have not created life itself. Science has never been able to breathe the magic of life into inanimate matter.

Nor has science been able to determine the likelihood that all this will happen. How often has nature succeeded in creating life out of nonlife? If nature performs that experiment on a thousand earthlike planets, does it succeed in every case, or once in a while, or hardly ever? Perhaps it succeeded only once, and we are the result of that rare occurrence.

Many scientists feel the answer is clear. They see life as a natural outcome of the laws of physics and chemistry. They point out that the earth is an undistinguished planet, made of elements found in abundance throughout the Universe. They note that the building blocks of life should also be abundant on every earthlike planet in the Universe. They are confident that wherever planets like the

119

earth are found—and evidence indicates that untold numbers exist in the known Universe—life will emerge.

How can we determine whether that is so? Is the creation of life out of nonlife an easy experiment for nature to perform, or a difficult one? Big stakes ride on the answer to that question, because if the creation of life is an easy experiment for nature, many planets in the Universe may bear life. But if the experiment is difficult and rarely succeeds, our fertile planet is a rare exception in an otherwise barren Cosmos.

The discovery of life anywhere else in our solar system would go far toward settling that question. Life on *one* planet—the earth—tells us nothing about the probability of life in the Universe, but life on *two* planets in one solar system would tell us nearly everything. For if life has arisen independently on two planets in a single solar system, it cannot be a rare and unlikely accident. If life is found on any other planet or moon in our own solar system, we will know that life is probably common in the Universe, and countless inhabited planets may exist on other stars. No scientific discovery more significant in its implications can be imagined.

Where in the solar system shall we look first?

Mercury is the closest planet to the sun, and seems like a good place to start. However, Mercury is not too promising as a potential abode of life. It is a small, baked planet, not much larger than the moon, and moonlike in appearance, with mountains, plains, and low-lying seas filled with dark lava. Mercury is a desolate world, battered by rocks from space, airless, waterless, and almost certainly lifeless. The sun is unbearably close, enormous—a scorching wheel of flame. It crawls across the sky; a day lasts three months on Mercury. During those long, hot afternoons, the temperature rises to 800 degrees Fahrenheit.

A FAMOUS EXPERIMENT. Stanley Miller, at the time a graduate student working for Harold Urey, performed an experiment at Urey's suggestion in 1953 in which he produced some of the molecular building blocks of life by passing an electric spark through a mixture of gases. According to current scientific ideas on the origin of life, life may have started on the earth when these molecular building blocks were produced in this way and then assembled by chance collisions into the first self-reproducing organisms.

121

Lead would melt in a puddle on the surface. The nights are bitter cold; at dusk, the thermometer plummets to 300 degrees below zero.

These extreme variations of temperature occur on Mercury because the planet lacks oceans and an atmosphere. The earth's temperatures do not vary so sharply between day and night because our oceans and atmosphere store up the sun's heat and distribute it over the face of the planet. But Mercury's closeness to the sun, and the great intensity of the sun's heat have driven off all its atmospheric gases and water. The planet must be a barren place, devoid of any form of life imaginable to us. Mercury would offer the human visitor vistas of stark beauty. However, the planet is unlikely ever to be visited by mankind.

MERCURY. *(opposite)* A close view of the barren, cratered, moonlike surface of the planet Mercury, photographed from a NASA spacecraft.

Venus, the next planet out from the sun beyond Mercury, was once thought to be a rather pleasant place. It has an atmosphere, fleecy white clouds, and should have a mild year-round climate, similar to the climate on the Caribbean islands. Heavy clouds conceal the surface of Venus from our view, but observers have always nourished the hope that these clouds drift over luxuriant forests, exotic fauna—and, perhaps, those Venusian inhabitants, described more than 100 years ago by the French astronomer Fontenelle as "a small dark people, burned by the sun, full of wit and fire, arranging festivals, dances, and tournaments every day. . . ."

We know today that this is not so. The clouds are droplets of deadly sulphuric acid, the temperature on the surface is greater than a thousand degrees Fahrenheit, the air is a dense, asphyxiating blanket of carbon dioxide. The breezes that stir the atmosphere bring no refreshment; they feel like the blast of a furnace. The pressure of the dense atmosphere is crushing—a force of 10,000 tons on the body. Venus is a hell hole. No flesh-and-blood creature could stand it.

The infernolike conditions on Venus were a great surprise to astronomers, because Venus and the earth are sister planets and should have similar climates. The two planets have approximately the same size and weight, are made of the same materials, and are only moderately different in their distances to the sun. Yet the earth is ideal for the evolution of life as we know it, while Venus is unimaginably hostile. Why is that so?

The answer is interesting, and has to do with the atmosphere of Venus. Every planet receives its warmth from the sun, whose rays are absorbed by its surface. If the planet has no atmosphere, it radiates most of the sun's heat back to space. However, if the planet has an atmosphere, the atmosphere tends to bottle up the heat from the surface and prevent it from escaping to space. As a result, the planet becomes considerably warmer than it would be if it were a naked, airless body of rock. This warming effect of a planet's atmosphere is called the *greenhouse effect*.

The earth's atmosphere also creates a greenhouse effect, but the effect is fairly small. The greenhouse effect produced by the Venus atmosphere is far greater, because the atmosphere on Venus is a hundred times denser than the earth's atmosphere. Moreover, the Venus atmosphere consists almost entirely of carbon dioxide, which is exceptionally effective in blocking the flow of heat from a planet's surface. This thick, insulating blanket of carbon dioxide explains the searing temperatures on the surface of Venus.

In spite of the hostile conditions on Venus, the Soviet Union succeeded in landing several packages of instruments on its surface during the 1960s and 1970s. These landings were a tour de force of planetary exploration. The Soviet landing craft carried out chemical tests and took photographs of the ground around their landing sites. The surface of Venus, never seen before, turned out to be a barren, gloomy landscape, the light level being about the same as that in a severe thunderstorm on the earth. The ground was strewn with rocks, presumably cast out in recent meteorite impacts. No sign of life—past or present— was visible.

Venus, like Mercury, may never be visited by man. The ship and its crew would have to be protected not only against the searing oven heat, but also against the pressure of the dense carbon dioxide atmosphere, which is more than one ton per square inch—equal to the pressure in the ocean at a depth of 3,000 feet under water. Venus may be the earth's sister planet, but it is not a tempting place for human colonization, or even for the briefest stay.

Beyond Venus, and beyond the earth, lies Mars, the red planet.

Conditions on Mars are far less hostile to life than on Venus, although not as comfortable as on the earth. During most of the Martian year the climate is extremely cold and dry. It resembles the climate in the Antarctic desert, but is even more severe. The atmosphere is very thin, the pressure on the ground being the same as the pressure in the earth's atmosphere at a height of 100,000 feet. The air on Mars consists mainly of carbon dioxide, as on Venus. However, the Martian blanket of carbon dioxide is too thin to produce much of a greenhouse effect.

In the search for extraterrestrial life, Mars stands out above all our planetary neighbors in importance because, although it is dry today, it seems to have had an abundance of water at an earlier time. Water is the quintessential ingredient for the emergence of life from nonliving matter. Water provides a fluid medium in which the molecular building blocks of life can collide again and again, to carry on the chemical reactions that make up the ongoing business of life. The basic molecules of life may exist in abundance on a planet, but unless they are dissolved in water so that repeated collisions can occur between each molecule and its neighbors, life cannot evolve.

Why do planetary scientists believe Mars had a great deal of water in the past? Their reasoning goes back to the conditions that prevailed at the beginning of the solar system, when the sun and planets were first condensing out of the parent cloud of the solar system. We know the parent cloud must have contained a substantial amount of water, since the constituents of water—hydrogen and oxygen—are relatively abundant throughout the Universe, and would have been present in the parent cloud as well. As the parent cloud cooled, the water vapor in the cloud must have condensed into grains of ice. These ice grains were then swept up into the bodies of the earth and Mars, and the other planets as well, when they collected out of the parent cloud.

At first, the water was trapped in the interior of each planet, but later it escaped to the surface. Geologists believe that nearly all the water in the earth's oceans came from the interior of our planet in this way, carried upward as bubbles of water vapor in molten rock.

Photographs taken at close range from NASA spacecraft reveal that Mars also has volcanoes—huge piles of congealed lava, some larger than any volcano on the earth. The Martian volcanoes are extinct today, but they were active in the past, spewing water vapor and other gases into the Martian atmosphere just as they did on the earth. In that early period Mars may have been nearly as wet as the earth.

Other NASA photographs show clearly that an abundance of water did, in fact, exist on Mars when the planet was younger. Channels that look like ancient riverbeds, some with midstream islands and branching tributaries, provide unmistakable signs that torrents of water once ran across the Martian surface. Geologists, studying thousands of photographs of Mars taken from NASA spacecraft orbiting the planet, have concluded that enough water may once have existed on Mars to cover the entire planet with a shallow sea 100 feet deep.

Current conditions on Mars prevent liquid water from appearing on the surface. All the water on the Martian surface today is in the form of ice, either at the poles or frozen into the soil as Martian permafrost. Because the Martian atmosphere is so thin, when the temperature rises the ice evaporates into the air instead of melting and forming a pool on the surface.

Yet there is evidence that conditions were more agreeable on Mars when it was a young planet. The climate was warmer then, and the atmosphere was probably denser. As a consequence, ice on the surface would melt instead of evaporating, forming rivers, lakes, and seas on the planet. At an earlier time, Mars may have been both warm and wet.

Any Martian life that evolved during these earlier periods of wetness and warmth may have adapted, by Darwin's slow process of natural selection, to the harsher conditions that came to prevail on the planet later on. The consensus among American scientists is that Mars has no life, but until we land on the planet, no one can be sure. It is possible that the descendants of early Martian organisms still survive on Mars today.

If that Martian life exists, it is not likely to be as advanced as life on the earth. The exceedingly dry conditions that developed on Mars later on must have slowed the pace of evolution there, and certainly would not have allowed the progression from simple to complex forms, and finally intelligence, that took place on the earth.

What kind of life would have appeared on Mars, before the pace of Martian evolution was diminished? The earth's history gives a hint of the answer. The severe Martian climate could have set in as early as several billion years ago. At that time, the highest forms of life on the earth were bacteria and one-celled plants. If evolution ran a parallel course on the two planets in their early years, bacteria and simple plants—or their fossilized remains— may be the most advanced signs of life we can hope to find on Mars today.

Yet the discovery of Martian organisms of any kind—
even as primitive as bacteria—would be of enormous interest. The existence of these forms of life would tell us at
once that nature has conducted independent experiments
on the evolution of life out of nonliving matter on two
planets, and on both planets the experiments have succeeded. That would suggest, in turn, that the creation of
life out of nonlife is not an exceedingly difficult matter for
nature to arrange, but something likely to happen anywhere in the Universe, when circumstances are favorable.

THE DESERTS OF MARS. This arid scene, photographed from the Viking Lander on the surface of Mars, is typical of many Martian landscapes. Mars is exceedingly dry today, with no liquid water on its surface and only a trace of water vapor in the atmosphere. It is also quite cold for most of the year and at most latitudes, the average global temperature being 40 degrees below zero Fahrenheit. However, the planet shows numerous signs of having been warmer and wetter in the past, and perhaps quite hospitable to life.

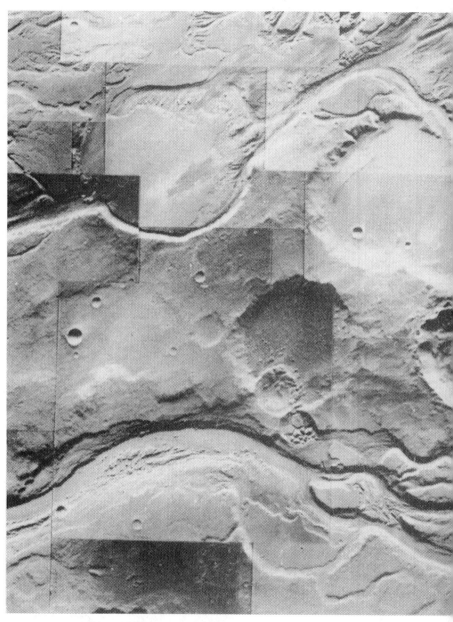

EVIDENCE FOR A WETTER MARTIAN CLIMATE. Many places on Mars show evidence of erosion by large amounts of water in the past. The braided channels near the bottom of the photograph,

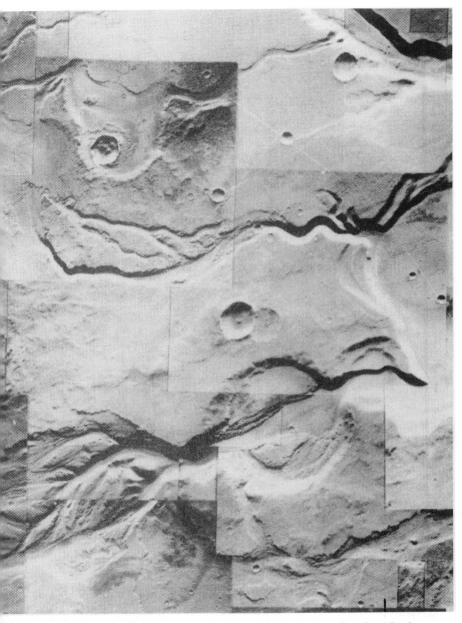

directly beneath two large craters, are an example; they look very much like the channels carved by rivers on the earth when they flow sluggishly across flat terrain.

VOLCANOES: SOURCE OF WATER ON AN-
CIENT MARS. Mars has many extinct volcanoes. The largest Martian volcano—and the largest
volcano in the solar system—is Mount Olympus,
a huge pile of congealed lava 70,000 feet high
and 300 miles across at the base.

On the earth, the eruption of gases from
volcanoes is believed to be the source of the
water in the oceans and the gases in the atmosphere. The water and gases, which had accumulated in the interior of the earth when our planet
first formed, remained there until they were carried up to the surface during volcanic eruptions
as bubbles of vapor in the molten rock. The
presence of many large volcanoes on Mars indicates that Mars could also have had large vol-
umes of water at one time. The source of the
water disappeared when the volcanoes became
extinct.

The size of a planet determines whether it
will be volcanically active and have oceans and
an atmosphere. Planets as large as the earth re-
tain their internal heat and remain volcanically
active for a long time. Mars is smaller than the
earth, and did not hold its heat as well as our
planet. That explains why the volcanoes it had
at one time are now extinct, while the earth's
volcanoes are still active.

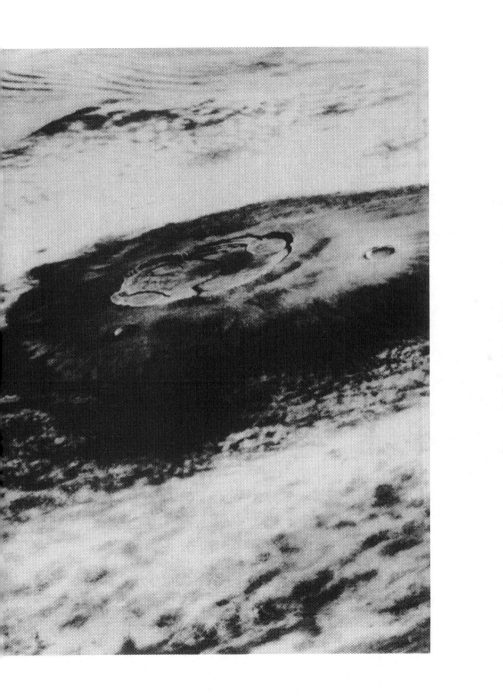

9

Mission to Mars

A spidery object dropped down onto the plains of Mars in the summer of 1976. Overhead circled its mother ship, an artificial Martian satellite created on the earth. The insectlike automaton on the ground beneath rested, checked its vital signs, and then began to go about the tasks that had brought it to the red planet.*

The automaton's gaze roved over the desert scene. Then a long trunk uncurled and picked up a handful of reddish soil. The automaton deposited the soil in its maw and began to digest it for signs of Martian life.

The information sent back to the earth by the automaton ignited a controversy that raged for a time and then subsided, but still smolders. Did the automaton find evidence of life on Mars? One experiment performed by the

*NASA called the trip to Mars the Viking Project. The automaton is known officially as the Viking Lander.

automaton seemed to say it did. The experiment tested the soil for the presence of Martian microbes, a simple form of life, but one whose presence would still give an affirmative answer to the question: Is the evolution of life so likely in the Cosmos that it could have occurred separately on two planets in one solar system?

How did the automaton test for Martian microbes? It added to the soil a solution with certain foods dissolved in it that microbes find appetizing. The foods, however, were specially prepared so that they contained a small amount of a radioactive substance—in particular, radioactive carbon.

The idea was that if there were Martian microbes in the soil, the microbes would eat the food, including the radioactive carbon, and then give off some of this radioactivity as a waste product, in the form of radioactive carbon dioxide. The microbes would eat the radioactive carbon in their food, for example, and exhale some of it as radioactive carbon dioxide. If detectors sensitive to radioactivity found that radioactive carbon dioxide was being given off by the Martian soil, that would indicate that microbes lived in it.

The experiment seemed to be completely successful. The Martian soil exhaled radioactive carbon dioxide, just as soils do in test runs of the experiment back on the earth, when the soils contain microbes.

Then the Martian soil was heated to a temperature above the boiling point of water, and the experiment was run again. If the soil really did contain microbes, the heat would destroy them and the radioactivity would disappear. The radioactivity disappeared, as expected. That check gave the experimenters more confidence that they really had detected microbes on Mars.

But other scientists disagreed. They pointed to another experiment performed by the automaton, which said with equal clarity that there was *no* life on Mars.

The second experiment did not search for life directly, but only for the molecular building blocks of life. These are known to chemists and biologists as organic molecules. If life existed on Mars, and even remotely resembled life on the earth, it would be made of these organic molecules. Even if the soil contained only the remains of dead and decomposed organisms, they would still show up in this test.

The results of the test for organic molecules were clear-cut. There were no organic molecules—no building blocks of life—in the Martian soil.

Without the building blocks of life, how could there be life? It seemed clear that the experiment that looked for radioactive carbon dioxide must have misled the scientists. Their experiment must have detected an obscure chemical reaction, instead of microbes.

But chemical reactions usually are not so sensitive to heat. If the microbe test was measuring a chemical reaction, and not life, why did the radioactivity disappear when the soil was heated?

Other checks were run, and they seemed to support the Martian microbe theory. But the scientists who opposed the idea of Martian life asked, "If there is life on Mars, where are the dead bodies? Where are the organic molecules?" This seemed to the space agency to be a telling argument. The official NASA conclusion was that no one could be certain of the matter, since the methods used in searching for life were indirect, but—there seemed to be no life on Mars.

Some American and Soviet scientists disagreed. The Soviet scientists pointed out that if Martian life exists, it is probably not to be found on the surface of the planet where Viking searched for it, but buried deep underground at the edge of the permafrost layer, where liquid water may exist. With this thought in mind, the Soviet scientists plan to drill at least 30 feet into the Martian soil when they conduct their own search for Martian life. Until this is done, one Soviet scientist said, "Viking results cannot be regarded as the last word."

The Soviet Union has become exceedingly interested in all aspects of the exploration of Mars in recent years, shifting its attention from an earlier preoccupation with the planet Venus. Surprisingly, Phobos, one of the moons of Mars, comes first in Soviet plans before the landing on Mars itself.* The reason for this is extraordinary: As much as 20 percent of Phobos may consist of water!

Water is hard to come by in space. It is not needed primarily for drinking, because in a manned mission the crew's waste water can be recycled and purified until it is drinkable. Water is important in space mainly because it provides a powerful rocket fuel. Of course, water itself will not burn in a rocket engine. But water is a compound of the two elements hydrogen and oxygen. If water is separated into these two gases, and the gases are then cooled and liquified, the liquid hydrogen and liquid oxygen that result make an excellent combination for propelling rockets—one of the best rocket fuels known.

A considerable amount of energy must be expended to break apart the water molecules and obtain the separate hydrogen and oxygen. However, the energy can be supplied by a small nuclear reactor. The nuclear reactor can be carried to Phobos on one of the first flights to the Martian moon. Once set up on the surface of Phobos, it will run for a very long time without additional fuel.

---

*Mars has two small moons named Phobos and Deimos after the horses that drew the chariot of the god of war.

After the hydrogen and oxygen gases have been produced, they must be cooled and condensed into liquids. (If they were left in the form of gases, they would occupy too much volume to be carried on board the rocket.) But the same nuclear reactor that separates the water into hydrogen and oxygen can also supply the electricity needed to refrigerate the hydrogen and oxygen gases until they are liquified.

Explorers of Mars gain a great advantage if they can pick up the fuel for their return trip—made from water in this way—at their destination, instead of carrying the fuel all the way from the earth. If a rocket ship starts out for Mars carrying the fuel it needs for the round trip, it pays a double penalty in weight. The ship has to carry not only the fuel that will be burned on the return trip to the earth from Mars; it must also carry the additional fuel needed to propel *that* cargo of fuel to rocket speeds when the ship leaves the earth at the start of the voyage.

For a trip to Mars and back without refueling, the weight of ship, supplies, and fuel might come to several million pounds. Lifting that much weight up from the surface of the earth would be a costly undertaking. Refueling at Phobos might cut the weight of a manned mission to Mars to half or a third of this amount, and make the trip less expensive.

It might seem at first that the water, and the rocket fuel that would be made from it, could be obtained on Mars without going to Phobos at all. Mars, after all, is also believed to have a considerable amount of water in frozen form under the surface. The disadvantage with that plan is that it is difficult to land on Mars and pick up the fuel, because of Mars' gravity. As the ship approaches Mars, a considerable amount of fuel must be consumed in resisting the pull of Martian gravity and slowing the spacecraft

141

down, so that it does not whip past the planet but settles into an orbit around it.

And after the main ship is in an orbit circling Mars, more fuel must be consumed by the small ship that will actually land on the planet, so that it makes a "soft" landing and does not crash into the surface. Then, after a supply of fuel has been picked up on Mars, fuel must be burned once more in taking off from the surface, again because the pull of Martian gravity must be countered. In fact, the fuel expended in landing on Mars and taking off again cancels much of the gain from producing the fuel on Mars in the first place.

But the water and fuel will be much easier to obtain from Mars' moon, Phobos. The main ship has to use some fuel in slowing down as it approaches the vicinity of Mars and its moons; otherwise, it would hurtle past both Mars and Phobos and go on into space. But once the ship has been slowed down enough to prevent that from happening, the next step—the actual landing on Phobos—is simple. Because Phobos is a tiny moon—15 miles in its longest dimension, about the size of the island of Manhattan—the pull of its gravity is so weak that a spaceship does not have to use an appreciable amount of rocket fuel to slow down for a soft landing; it merely hovers over the surface, blowing gently on the ground below.

And because Phobos' gravity is so weak, it takes hardly any rocket power to blast off from the little moon again, after you have landed there and picked up fuel and water. Humanpower is sufficient; a person could leap off the surface of Phobos and go into space with one good running jump.

The best procedure for the Mars explorers would be to stop at Phobos before going on to the planet, to pick up the fuel needed for the Mars landing itself. At that time, they can also obtain the fuel needed to take off from Mars after the exploration is completed and to return to Phobos. Once back on Phobos, they can refuel again for the main voyage back to the earth and the descent to the earth's surface.

Phobos is uniquely valuable to the explorer of the solar system. It can be a fuel and water resource, not only for the exploration of Mars, but even for the support of a scientific outpost on our own moon. In fact, it would take less rocket fuel, and cost less, to bring water to our moon from Phobos, than it would to bring that water directly up to the moon from the earth.

That seems surprising at first, because the distance from the earth to Phobos is hundreds of times greater than the distance from the earth to our moon. However, a cargo ship carrying water from the earth to the moon must consume a very large amount of rocket fuel in breaking loose from the powerful grip of the earth's gravity, whereas a cargo ship carrying water from Phobos to the moon does not have to use any fuel to break away from Phobos' gravity. The ship still has to break loose from the gravity on Mars, but that is less than half as powerful as the force of gravity on the earth.

These interesting possibilities depend on the assumption that Phobos really contains a large amount of water. Planetary scientists think it does, because in some important respects Phobos resembles certain kinds of meteorites called *carbonaceous chondrites*—pieces of planetary matter from the asteroid belt—that have a water content of as much as 20 percent. Another indication of water on Phobos is a set of grooves in the surface of the moonlet that look like places where steam escaped following a collision between Phobos and an asteroid.

A visit to Phobos is high on the list of U.S. and U.S.S.R. priorities for future Mars missions. Meanwhile

143

the U.S.S.R. has firm plans for a series of visits to Mars itself, starting in the mid-1990s, when a Soviet spacecraft will drop into an orbit around the planet to become an artificial Martian satellite. The spacecraft will reconnoiter Mars from orbit. It will also release a large balloon in the Martian atmosphere. Inflated with helium, the balloon will float in the thin air of Mars, rising to a height of about three miles during the day, and moving with the circulation of the winds. It will trail a container of instruments—a flexible snake-like sheath about ten feet long and six inches wide—including a camera as well as equipment for detecting water and measuring the chemical composition of the soil.

As night approaches and the temperature drops, the balloon descends until its payload of scientific instruments rests on the ground. When the rays of the morning sun strike the balloon, the gas inside expands and the balloon rises to drift perhaps three hundred miles to another site, before descending again and settling in for the night.*

A few years later, Soviet scientists plan to deposit a small, driverless automobile on the surface of Mars to wander over the Martian surface. The rover is likely to be a six-wheeled vehicle, with oversized tires for coping with the rough Mars terrain. It will be steered by an electronic brain that has been instructed beforehand in the nature of the hazards that probably await it, and the best stratagems for surviving them.

These formidable hazards include a rock-strewn terrain and massive Martian dunes. If the small rover succeeds in

---

*The "balloon" is actually two balloons, one filled with helium and sealed, and the other is open to the Mars atmosphere. The open balloon is black and absorbs the heat of the sun, providing the extra buoyancy every morning that enables the two balloons to take off on their daily flights.

meeting those challenges, a much larger Soviet rover will be deposited on the surface of Mars, capable of traveling hundreds of miles. This rover will weigh three quarters of a ton—about as much as a small automobile. It will also be an automaton, moving around and performing its scientific tasks under the direction of an electronic brain.

However, the brain of the large rover will be charged with a new responsibility of the highest importance. It will attempt, for the first time, to collect samples of Martian soil from widely scattered locations, and send them back to the earth for study. That staggeringly difficult feat, if accomplished, will be a watershed event in the history of Mars exploration, for only then, at last, may the question of Martian life be settled.

Meanwhile, preparations for manned flights to Mars will be underway in the United States and the U.S.S.R. The manned exploration of Mars may begin with a manned interplanetary loop around the planet and a return to the earth without landing. That tests the reliability of the spaceship on the long interplanetary journey, before the space travelers contend with the additional complexities of the actual descent to the surface of the planet. The United States followed this conservative, two-step plan in the moon landing project.

If successful, the pioneering manned flight around Mars may be followed by a landing—the first landing of men and women on another planet—in the early decades of the twenty-first century. But such a flight would mean a stay of perhaps two years away from the earth for the crew of the mission. Manned flights in space of such long duration present special problems for human survival that may turn out to be insoluble.

145

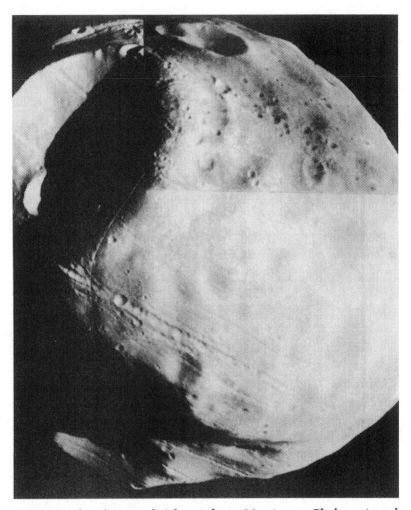

PHOBOS. This photograph *(above)* shows Mars' moon Phobos, viewed from the *Viking* spacecraft in 1977 at a distance of 380 miles. Phobos is named after one of the horses that drew the chariot of the god of war. It is a small moon, about 12 by 16 miles in size, and was probably captured from the nearby asteroid belt.

Phobos is made of a low-density material similar to the material in a type of meteorite called a carbonaceous chondrite, which may contain as much as 20 percent water. If Phobos contains an abundance of water, it will become an oasis in the desert of space, invaluable in future missions to Mars and the other planets.

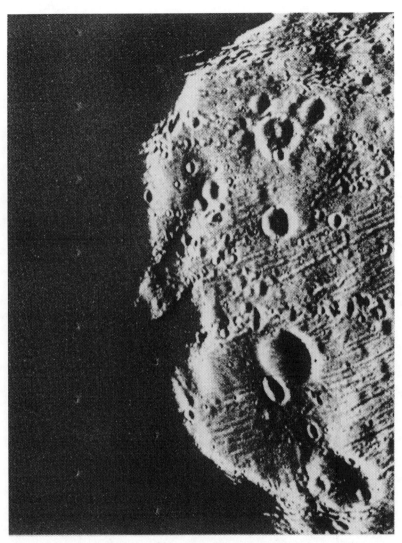

A closer view of Phobos from 200 miles reveals many mysterious grooves *(above)*. These may be fractures in the body of Phobos caused by the impact of large asteroids with the moon. They could also be caused by Phobos' closeness to Mars. Mars' gravity, pulling at the near side of Phobos more strongly than the far side, could be breaking up the satellite. In the future, Phobos may crash into Mars; or the Martian gravity may tear it into fragments that circle Mars like the rings of Saturn.

10

The Martian Experience

A good many years ago I took part in a small meeting at the NASA research center on Wallops Island in Virginia at which several scientists presented their visions of the future. Neil Armstrong, first man to walk on the moon, was there. Homer Newell, my mentor in NASA, was there. Some famous futurists were present, notably Arthur C. Clarke and Wernher von Braun.

After all this time, I still remember the vivid impression created by Dr. von Braun's suggestion that a colony on the moon could obtain oxygen by crushing lunar rocks. This was the first time I had heard of that idea, and it seemed like a very audacious proposal.

Of course, rocks do contain a great deal of oxygen; it makes up about half their weight. In fact, half the mass of the entire moon is oxygen. However, the oxygen in rocks is not readily available for breathing because it is tied very tightly to the other elements in the rocks. Von Braun's thought was that if you crushed the lunar rocks you could extract the oxygen from the crushed rock powder by heating it and breaking the chemical bonds. Still, obtaining oxygen from rocks seemed almost a miracle—like Moses striking a rock in the desert to obtain water for the Israelites.

After you have oxygen, von Braun went on to say, you could bring in liquid hydrogen from the earth and combine it with the oxygen to obtain water. Most of the weight of water resides in the oxygen atoms it contains; in a pound of water, for example, the hydrogen atoms only weigh four ounces, and the oxygen atoms make up the remainder of the pound. Bringing material from the earth to the moon is expensive—about $50,000 a pound—but if you have to bring only the relatively light hydrogen atoms to the moon, and the heavy oxygen atoms are already there, you have an inexpensive way of making water for the lunar colony.

Von Braun also talked about exploring Mars. That has been looked at carefully in the last few years by scientists in NASA and other organizations, and their conclusions are not too different from his proposal. According to one plan, two ships will make the trip, each carrying six men and women, including a physician. The ships fly buddy system; each has provisions for 12, and if one ship is disabled, its crew can transfer to the other for a safe return to the earth.

Each ship is 200 feet long and weighs 600 tons, of which more than two thirds is fuel. All the fuel needed for the round trip is on-board. Later, when facilities for production of fuel have been set up on Phobos, that will not be necessary. This is a pioneering flight, and no amenities await the crew at their destination.

The ships have been assembled in orbit at the space station, during the year before the flight, from units carried up, 75 tons at a time, by large rockets. Each Mars ship dwarfs the space station itself.

Conference participants and staff listen to Wernher von Braun at the NASA meeting on the future of space, as he explains how manned missions to Mars might be accomplished.

During the previous week the ships have been fueled with liquid hydrogen and liquid oxygen. Now they circle in orbit awaiting takeoff. Their engines ignite in a roaring inferno. Jets of hot gas, rushing out of the rocket engines at a speed of 28,000 feet per second, push back on the engines as they escape, driving the ships forward. The nine-month journey to Mars has begun.

Three days out, the crews look back on the earth, now more than three million miles away. It is a tiny, jewel-like sphere of blue, brushed with green and gold. The ships fly in close formation, each crew fully occupied with space-craft systems checks. Then, with the ships checked out for the long flight across the solar system, the commanders undertake a delicate maneuver vital to the success of the mission. They will turn the two ships so they are stern to stern, and connect them by a 600-foot tether, so that they form a single craft 1,000 feet long.

Slowly the ships back toward one another, feeling for the adapters. They dock, and connect the two ends of the tether. Then the two ships back off cautiously, and the tether unreels until the crew quarters, located at the far ends of the ships, are 1,000 feet apart.

The operation is successful. Small rockets, oriented at right angles to the axis of the oddly shaped craft, set it into a lazy spin at two revolutions per minute. The crew, located at opposite ends of the 1,000-foot structure, feel a centrifugal force that pushes them back against the outer walls of their cabin. Prior to that moment they had been in zero gravity. Now they feel the force of an artificial gravity.

The force is reassuring. It keeps their innards in place; more important, it prevents vertigo—extreme, incapacitating dizziness and nausea—as well as a serious deterioration of the bone and muscle that sets in when the body is weightless.

In the first days of the flight, when the ships were still flying in zero gravity, both crews were afflicted by dizziness and nausea. The inner ear was the principal cause of their problem. The inner ear contains tiny particles of calcium, suspended in a fluid. These calcium particles press up against sensitive hair cells within the ear that are connected, in turn, to the brain. Every time the body changes its position, the particles press against different hair cells, which send a message to the brain telling it what the body is doing—how it is turning and moving.

Normally, when the possessor of that inner ear is on the earth, the force of the earth's gravity controls exactly how the tiny particles press against the hair cells in various positions of the body. The circuits of the brain that interpret the messages from the hair cells are wired so that they draw correct conclusions about what the body is doing when gravity is the controlling force.

But in the absence of gravity, the particles press against different hair cells, which send spurious messages to the brain that no longer give it an accurate picture of the body's movements. These signals from the inner ear may conflict with the other evidence the brain receives from the senses. For example, when the head is turned or tilted, the eyes tell the brain exactly how the head has moved; but if a person is in zero gravity, the brain gets a different and conflicting message from the hair cells in the inner ear. The brain, whose circuits evolved a long time ago in response to the needs of an animal living on the surface of the earth, cannot make head nor tail of this contradictory information. The result is vertigo—a debilitating dizziness and nausea.

The deterioration of the body caused by weightlessness is much more serious than the vertigo. In fact, it can be lethal. For one thing, calcium dissolves out of the bones during an extended stay in space and the skeleton becomes porous and brittle. The experience of the American and Soviet astronauts suggests that 30 percent of the material in the bones of the crew could disappear during a round trip to Mars in zero gravity, leaving their skeletons in a dangerously fragile condition.

The weight-bearing bones of the body—legs, heels and toes, and backbone—are the most affected. The brittleness of bones is less serious during the space flight itself, because the crew members are weightless then, and their bones are subjected to less stress than normal. The biggest problems arise when the ship returns to earth. Then, as the crew members emerge and their bodies feel the full force of the earth's gravity once more, a broken spine and paralysis may result. While the bone loss can be reduced by vigorous exercise, Soviet experience with long stays in space suggests that it is not entirely eliminated.

The calcium that dissolves out of the bones creates another major medical problem for the Mars crew. The calcium enters the bloodstream and then is filtered out of the blood by the kidneys. However, the continued passage of excess calcium through the kidneys may produce kidney stones—a painful ailment, possibly requiring major surgery. Even assuming a qualified surgeon on board, surgery is difficult in zero gravity, because organs tend to float out of the body cavity, and a mist of blood droplets fills the cabin.

Still other medical problems arise if the trip to Mars is made in zero gravity. Muscle, as well as bone, wastes away, as it does when people stay in bed for a long time during an illness. One would expect the leg muscles to be particularly affected because there is not much work for the legs to do in the absence of gravity. A Soviet cosmonaut lost 15 percent of the muscle volume in his legs during an 11-month stay in space in 1987.

But, surprisingly, the muscles of the arms also atrophy in zero gravity. That means degraded performance by the crew in critical tasks—both heavy work, using the hammer grip of the arm and hand, and delicate operations, using the precision grip of thumb and fingers.

The heart also atrophies. The heart muscles of Soviet and American astronauts shrank as much as 20 percent during long stays in orbit. That is not surprising, for if muscles are not exercised, they shrink. In a weightless condition, the heart has less work to do, and, like any other muscle that is not used, it gets smaller and weaker. When the crew lands on Mars after the trip across the solar system, or when they return to earth, their weakened hearts may not be able to stand the added strain of pumping blood against the force of gravity. Sudden cardiac arrest may follow.

All these health hazards are avoided at the outset by connecting the two Mars ships and spinning them around their center to create an artificial force of gravity. How much artificial gravity is needed? If the cabins for the crew are at the far end of the spinning dumbbell, a thousand feet apart, and the whole system spins at two revolutions per minute, the artificial force of gravity works out to be two thirds the force of gravity on the earth. That amount of artificial gravity is a good way to prepare the crew for the stay on Mars, because it happens to be midway between the force of gravity on the earth and on Mars.

While artificial gravity protects the crew from the effects of weightlessness during the long trip to Mars, other dangers still confront them. The most serious hazard is radiation poisoning. The ships themselves are not radioactive, but the space through which they move is filled with a life-threatening intensity of cosmic rays. Cosmic rays are extremely energetic particles that come mainly from other parts of the galaxy. The rays damage the cells of the body by ejecting electrons from atoms in the cells and by disrupting atomic nuclei. The cells of the reproductive organs, the bone marrow, and the eyes are particularly vulnerable. Infertility, cancer, and cataracts are common consequences of radiation damage.

The harmful cosmic rays are hundreds of times more intense in space than they are on the surface of the earth. Life on the surface of the earth is shielded from cosmic rays by our planet's atmosphere, which absorbs the dangerous radiation, and by the earth's magnetic field, which turns the rays aside. These natural shields are missing in space travel. As a result, during a single trip to Mars the members of the crew receive as much harmful radiation as they would normally receive in an entire lifetime on the earth.

The crew can be protected against cosmic rays by a thick layer of material added to the outer walls of their cabin, as a substitute for the protection provided to life on the surface of the earth by our atmosphere. However, hundreds of tons of shielding are necessary. It is more practical to let the crew make the journey without special protection, and limit each crew member to one trip to Mars in his or her lifetime. Although that doubles the lifetime dose of radiation received by each crew member, the increase is no more than is experienced by a person in moving from a city at sea level to a city at an altitude of 5,000 feet—from New York to Denver, for example.

But cosmic rays are not the only dangerous radiation the crew encounters in its trip. Now and then, the surface of the sun erupts in a violent outburst, called a solar flare, that sends a cloud of fast-moving and potentially lethal particles through the solar system. These solar eruptions, which cannot be predicted, last for up to 24 hours and produce radiation millions of times more intense than we experience on the surface of the earth. If a strong solar flare erupts during the Mars trip, the crew of the Mars ship can receive a fatal dose of radiation in less than an hour. No danger encountered on the journey is more serious than this.

A heavy shield around the crew quarters would give the crew some protection from solar flares, but again at the cost of hundreds of tons of added weight. A better solution is to equip the ship with a "storm cellar"—a small enclosure, heavily shielded from radiation, within which the crew can remain for the duration of the flare. A storm cellar sheltering a crew of six would require no more than 10 tons of shielding.

However, the crew must move quickly when a flare occurs. If they wait until the energetic particles from the flare actually arrive at the ship before crawling into the cellar, they are likely to suffer serious injury. Fortunately, solar flares usually give a warning signal—a burst of relatively harmless radio waves that travels through the solar system at the speed of light, faster than the deadly particles in the flare, and arrives about an hour before the particles. If the crew is on the alert for these radio bursts, they will have an hour's warning—sufficient time to reach the storm cellar. But crew members working outside the ship are in danger. They may receive a lethal dose of radiation before they can get back inside.

•

The tethered ships move silently through space, rotating majestically. No eye perceives the impressive sight. Small rockets puff gently, keeping the rate of rotation of the ships constant against the disturbing movements of the crew as they go about their duties. Instruments scan the surface of the sun continually for early signs of a dangerous flare. Life settles into the shipboard routine of systems checks, navigational fixes, eating, sleeping, and study.

The crews live and work in cylindrical modules 33 feet wide and 40 feet long, containing four levels. One module accommodates six men, but could accommodate 12 if required. Each man consumes two pounds of oxygen and two pounds of food daily, plus eight quarts of water for cooking, drinking, and washing. Six and a half quarts are recovered by daily recycling of wastes and condensation of atmospheric moisture. The remaining one and a half quarts come from the on-board water supply. Twenty tons of oxygen, food, and water will be consumed by the crew of each ship during the journey, but double that amount is carried as a margin of safety.

The psychological strains of the voyage are severe, but the crew have been tested for their stability as a team by long tours of duty in the space station. The micro-society of the ships is self-governing. Issues of daily living that do not involve life-threatening matters are decided democratically by consensus. In emergencies the authority of the commander is accepted and the forms of government revert to a military chain of command.

The Mars ships are complicated organisms, and ministering to their electronic and mechanical ailments requires vast technical knowledge. In the days of Apollo, hundreds of experts hunched in front of TV computer screens in "Mission Control," on the ground back in Houston, monitoring the performance of the spacecraft, initiating corrective actions, and using the astronauts as extensions of their fingers and brains. That was feasible then, because it took less than five seconds for radio signals to pass from Houston to the Apollo spacecraft and back. But the exchange of radio messages between the Mars ship and the earth can take as long as an hour. If the ship's instruments detect an explosive fuel leak or an overheated circuit, instructions from Houston for the "fix" will arrive too late to avert catastrophe.

Allowance has been made for that delay. An elaborate system of sensing instruments is imbedded in the ship's vital organs, and the vast knowledge of Houston's engineers is stored in its electronic brain. The ship has absorbed the wisdom of the experts; it forgets nothing, and learns by experience. Powerful on-board computers understand every circuit, can test every reflex and every procedure, diagnose every electronic ailment, and offer a cure. The ship is alive. It senses and reasons.

•

The two ships coast to Mars, arriving in its vicinity nine months later. While still more than a million miles away, the ships separate and approach the planet, flying in close formation once more. Gradually they accelerate under the pull of Mars' gravity. As they whip past the planet at 7,000 miles an hour, a short burn of the rockets deflects their course and they descend into the Martian atmosphere. The resistance of the atmosphere slows the ships further, and they settle into parking orbits.

Aerobraking is the term for this maneuver that uses the atmosphere to slow the ships. It is one of the most delicate and hazardous operations in the entire journey. Descend into the atmosphere at too steep an angle, and ship and crew burn up in the atmosphere. Descend at too shallow an angle, and the ship skips out of the atmosphere again, like a flat stone skimming over the surface of the water, and flies off into space.

When the aerobraking maneuver has been completed, the crews of the two ships turn their attention to the landing itself. Within each ship, three crew members have been designated for the descent to the Martian surface. The others stand by. The chosen crew members enter the Mars Excursion Module and begin their preparations for the landing. The Mars Excursion Module is an enlarged two-decked version of the Lunar Excursion Module used in the Apollo landings on the moon. It has been stowed aft of the crew quarters during the outward journey. Now, with the first landing party aboard, the Excursion Module disengages from its moorings and pushes gently free of the Mars ship.

Again using the resistance of the Martian atmosphere, the landing party aerobrakes the Excursion Module to take it out of orbit. At an altitude of three miles, the parachutes open—first the drogue, at a speed of 700 miles per hour,

159

and then the main chute—and the Excursion Module slows to 300 miles an hour. A thousand feet above the surface, retrorockets fire to lower the craft to a soft landing. The hatch of the Excursion Module opens and the crew members bounce to the ground, moving easily in the lighter-than-earth gravity of Mars. The first manned exploration of an alien planet has begun.

IV
A Journey into the Future

11

Encounter with a Star

One of the most remarkable of all findings in astronomy is the discovery that the Universe was already ancient by the time the sun and its family of planets came into existence. More than half the stars in the Universe are billions of years older than the sun. Earthlike planets circle around many of those older stars. Intelligent beings may have appeared on some of these earthlike planets a billion years or more before mankind appeared on the earth.

What will those intelligent beings be like? What will our own descendants be like in a billion years? The first humans appeared on the earth somewhat more than a million years ago; modern man appeared less than 50,000 years ago; and many of the inventions on which life depends today are less than 200 years old. Yet these intervals are the blink of an eye in the lifetime of a planet or star. The changes that can occur in a billion years are beyond our imagination.

Consider the march of new inventions in the last century: the telephone, automobile, airplane, then—at a quickening pace—radio, television, remarkable new medicines, the computer. . . . Each adds another touch of magic to human life. Contemplate another *billion* years of accelerating progress. These older civilizations may have surpassed our achievements long ago. We can expect they will have mastered the techniques of radio communication from star to star, and harnessed the energy required for a voyage across the galaxy, with greater skill than our scientists can hope to achieve for many centuries to come. Just as Columbus discovered the Indians, rather than the other way around, we must expect that these advanced societies will make their presence known to us before we are able to reach out to them.

Perhaps the advanced beings form a galactic network of intelligence, welcoming the newcomers one by one, as they cross the threshold of radio communication. No doubt our first contact with older races will be by radio, but eventually the desire to see these distant places and "people" may become overwhelming. Curiosity may drive us across the boundary of the solar system.

Yet an enormous gulf separates us from our nearest stellar neighbors. In the opinion of many scientists, the gulf can never be bridged. The closest star to the sun—Alpha Centauri, visible only in the southern hemisphere—is 25 *trillion* miles away. If we traveled at the speed of light—186,000 miles a second—we would need more than four years to reach it. At conventional rocket speeds the journey would take 50,000 years. Most of the other hundred billion or so stars in the Milky Way Galaxy—the huge cluster of stars to which the sun belongs—are hundreds or thousands of times more distant than Alpha Centauri. Is there no way we can reach them?

Despite the pessimism of scientists, several methods have been discussed. None is feasible today; each demands an extension of science beyond the limits of today's knowledge, or dramatic changes in human behavior patterns.

The first possibility requires a major advance in medical knowledge. If research can conquer the problem of suspended animation for human beings, space travelers may pass hundreds or thousands of years in frozen, ageless sleep, drifting across the Cosmos in an automated spacecraft. When the craft approached its destination, the travelers would be awakened by a computer. Rising, fresh as defrosted broccoli, they would undertake their tasks of exploration. After completing their survey, they would head back to earth, sleeping again until they were awakened to step out into a world bearing little resemblance to the one they left hundreds of centuries earlier.

Another possibility—terminal germination—is closer to the realm of current medical practice. It may be possible to transport germ cells—sperma and ova—which could be combined in artificial wombs as the craft approached the alien solar system toward the end of its journey. Raised in the care of computer parents, the crew members would step out full-grown to explore the new planets on arrival.

As a medical experiment, the fertilization of a mammalian ovum and the nurturing of the embryo up to the point of birth may be no more than 50 years off. Experiments along these lines with lower animals are a part of the medical revolution now in progress. It seems possible that this research will open an avenue to interstellar travel long before the problems of suspended animation of humans are overcome.

Still another possibility would depend on advances in the physical sciences. With radical improvements in rocket engines and fuel, an interstellar ship could be accelerated

165

to speeds approaching the speed of light. If the ship accelerated to two tenths of the speed of light, the trip to Alpha Centauri and back could be made in 45 years—within the lifetimes of the crew if they were young at the start.

But no rockets capable of approaching the speed of light are available at the present time. The rocket engine that comes closest to this goal is one powered by the energy of nuclear fusion—the same energy that powers the explosion of a hydrogen bomb. The nuclear fusion rocket does not exist yet, but it is not very far beyond the reach of today's technology.

Even with this advanced rocket, a ship weighing 10,000 tons—about the minimum needed for a 45-year round trip in space—would require 100 million tons of nuclear fuel to boost its speed to two-tenths of the speed of light. That includes not only the fuel used in accelerating the ship to this high speed, but also the fuel needed to slow it down on its arrival at the star, and the additional fuel needed to speed it up and slow it down again on the return trip to earth. The cost of the trip would be equal to the output of the world's economy for a century. These are not very attractive prospects.

The nuclear rocket is not the ultimate in high-speed travel. The ultimate could be achieved if all the fuel in a rocket were converted to pure radiant energy in accordance with Einstein's formula, $E = mc^2$. The fuel needed for that purpose is antimatter—large quantities of it.

Antimatter is the exotic counterpart to the ordinary matter that makes up the familiar Universe. We know from laboratory experiments that when equal amounts of matter and antimatter are brought into close contact, they annihilate one another in a great burst of energy. Antimatter used as a rocket fuel is hundreds of times more powerful, pound for pound, than the fuel used in nuclear fusion.

If copious quantities of antimatter could be manufactured on the earth and fed into an engine at a controlled, steady rate, we would have a rocket capable of approaching the speed of light.

A 10,000-ton ship powered by an antimatter engine could be accelerated to two tenths of the speed of light using only 5,000 tons of fuel. That would be an improvement over the 100 million tons of fuel required to achieve this speed with a nuclear-powered rocket. It could be accelerated to an even higher speed—99 percent of the speed of light—with 140,000 tons of fuel. At speeds as high as this, the ship could make the round trip to Alpha Centauri in less than ten years.

If such an engine ever became a reality, it would carry a bonus. Einstein's theory of relativity predicts that space travelers moving at speeds close to the speed of light will age less rapidly than their fellow humans who stayed behind on the earth, because time itself slows down at high speeds. Travelers making a trip to a distant star at very high speeds may return little older than they were when they departed, according to the theory of relativity. If the journey is long enough, the space travelers may find, on their return, that they are younger than their children.

Parents younger than their children? Physicists rarely make predictions as strange as this. Einstein's prediction that time slows down at high speeds seems so odd that for a time some scientists contested it vigorously. However, the effect has been tested in a number of experiments and found to be exactly as Einstein had said.*

---

*In one famous experiment performed in 1977, a very accurate atomic clock was flown around the world on a jet airliner. When it returned, the clock was compared to a sister clock that had stayed on the ground and was found to be running slow by just the amount Einstein's theory had predicted.

Suppose that with the aid of the hypothetical matter–antimatter engine, a ship has been accelerated to 99 percent of the speed of light for a trip to a star 25 light-years away. (A light-year, the astronomer's unit of length, is the distance light travels in one year. It works out to be slightly less than six trillion miles. If a star is 25 light-years away, an object moving at the speed of light takes 25 years to reach it.) If the ship travels at 99 percent of the speed of light, the round trip to that star lasts a little more than 50 years. When the space travelers return to the earth, all their friends are 50 years older, and many are dead.

But according to Einstein's theory, during a trip at 99 percent of the speed of light, the space travelers age only one tenth as rapidly as their friends on the earth. Fifty years have passed on the earth since their departure, but the travelers are only five years older than they were when they departed on their journey.

The theory of relativity predicts that time slows down for the traveler at any speed, and not only at speeds near the speed of light. For example, passengers on an airliner traveling from New York to Los Angeles are a few millionths of a second younger on arrival than they would have been if they stayed at home. But the effect only becomes striking at speeds close to the speed of light. At the speed of light itself, according to the theory, time comes to a complete stop. The crew of an interstellar ship, traveling at the speed of light, would reach their destination literally in no time, and would not age at all during their journey.*

---

*But, according to the theory, attaining the full speed of light is impossible for a spaceship or any other material object, because it would require an infinite amount of energy. The amount of energy required increases rapidly as the speed of light is approached. For example, accelerating to 99 percent of the speed of light takes ten times more energy than the acceleration to 90 percent.

The space travelers gain these benefits from the theory of relativity only if science can build an antimatter engine to carry them very close to the speed of light. Unfortunately, the construction of the antimatter engine, so common in science fiction, is fraught with technical difficulties. At present, we can only make minute quantities of antimatter, less than a trillionth of an ounce, in the laboratory; how will we ever make thousands of tons? And if we make tons of antimatter, in what kind of vessel will we store it? Any container made of ordinary matter would be annihilated by its contents. Yet, what seems beyond the reach of science at our level may be within the grasp of older and more advanced civilizations. An advanced society's science is a primitive society's magic.

A trip through space at a speed close to the speed of light presents other problems. Perhaps most serious are the collisions between the fast-moving ship and the atomic particles that drift through space—protons, neutrons, electrons, and the like. These particles become dangerous projectiles when a ship moves through them at high speeds. Each atomic particle penetrating the walls of the spaceship leaves behind a lethal trail of radioactivity. Bits of solid matter—tiny grains of ice and fluffy particles of dust less than a thousandth of an inch in diameter—are also scattered throughout the space between the stars. Each bit of matter, striking the hull of the ship at 99 percent of the speed of light, explodes with the energy of a small atomic bomb. A shield of lead at least 30 feet thick and weighing a million tons would be required to protect the ship against these hazards.

Perhaps elsewhere in the Universe scientists exist who have overcome these formidable problems. But from the perspective of humans, the antimatter engine seems so far beyond the limits of current technology that its realization must be regarded as thousands, if not tens of thousands, of years in the future.

Yet the stars beckon, and other ways of reaching them have been suggested. One proposal avoids the perils of high speed travel, but at a price that few people alive today would be prepared to pay. At the present rate of progress in rocket technology, in a few hundred years we should be able to build giant space arks, the size of the *Queen Mary*, which would be self-sustaining and able to maintain life for an indefinite period. These arks would be miniature earths, growing their own food and carrying with them the culture of the parent civilization. Generations would be born, evolve, and die on the way to a distant star.

The ark would replenish its supplies periodically by "fuel stops" at a succession of stars en route to its destination. Encounters with rogue comets would add to the resources of the space ark. There is evidence that the space-between the stars may be strewn with comets that have been torn out of their solar systems by an encounter with another star. These porous lumps of ice and rocky material will be treasure troves for the space traveler. The bits of rock they contain can be processed into aluminum, iron, and other materials for construction; their ice, melted, yields precious water for the inhabitants of the space ark; the water, broken into its components, hydrogen and oxygen, provides air to breathe and fuel for the ship's generators. And the water also yields deuterium, a fuel used in nuclear rockets.

Life would be full for everyone in the space ark, and each person would have a demanding task to perform in this microsociety. The survival of the ark would depend on qualities that are held in great esteem by society today. Emotional stability, tact, and a high level of cooperation would be well regarded in the confined world of the space ark.

Would anyone be eager to accept such missions, knowing he or she would never see the earth again? Judging by the past, there is little doubt that people would. Man is an exploring animal. From the ancient Greeks and the Polynesians in their seafaring canoes to the American pioneers, the history of exploration has been in large measure a history of men and women who struck out on dangerous and arduous journeys to settle unknown worlds, with no expectation of ever returning to their homes.

12

A Message

Since the early 1960s, television programs have been spreading out into space from the earth at a million-watt level. Today the total power radiated by the television stations of the world is more than one billion watts. In the course of the last 15 years, that expanding shell of television broadcasts, moving away from the earth at the speed of light, has swept past hundreds of stars. Old Jack Paar and Johnny Carson programs have carried the message to these stars that intelligent life exists on this planet.

Powerful radar beams, sprayed into space from the huge radars placed on the borders of the United States and the USSR to give warning of air and missile attacks, make the presence of humans even more evident to intelligent beings observing our planet from distant stars.

These signals cause the earth to shine a thousand times more brightly than the sun at frequencies in the television and FM bands. Radio astronomers in other solar systems, turning their antennae in our direction, will notice the outpouring of energy and realize that a scientifically advanced society exists on this star.

Our television signals tell the extraterrestrials more than that. From subtle shifts in the frequency of the signals as the earth rotates, they can infer the distance from the earth to the sun, the likely temperature on the surface of our planet, and, from that, what kind of life may exist here.

If astronomers in other solar systems have been monitoring our progress, they now have evidence that this life has crossed an important technological threshold—the threshold of radio communication. The extraterrestrial scientists can infer from their own experience that this achievement must be followed soon after by the mastery of travel in space—first from planet to planet, and then, not too long after that, by voyages beyond the boundaries of our solar system. Unwittingly, we have sent word abroad that we are preparing to enter the galactic community.

If those extraterrestrial beings exist, this may be the signal they have been waiting for. A message of welcome, beamed toward the sun, may be on its way back to us.

From what stars would the messages come? The civilizations that send them must not be too far away; if they are too distant, the signals that reveal our presence will not have reached them. Or, if the signals have reached these stars, their replies may not have had time to get back to us.

Any star within a dozen light-years from the sun is at a suitable distance. Suppose the star is just 12 light-years away. That means a Johnny Carson program leaving the earth in 1965, say, would get to this star in 1977 after a 12-year journey through space. If the reply went out promptly, it would take 12 years to get back to the earth, and would reach us in 1989. We should have heard from that star by now.

But the stars that are likely to commence a dialogue with the earth must not only be at the right distance; they must also be of the right age—neither too young nor too old.

This is very important. If a star is too young—if, for example, it is only one or two billion years old—life on that star probably has not had time to evolve to an intelligent level. After all, the earth existed for more than four billion years before a significant degree of intelligence appeared in the creatures on its surface. On planets circling stars a mere billion years younger than the sun, the summit of creation may be occupied by soft-bodied creatures similar to the worm and the jellyfish—for these simple animals were the highest forms of life on our planet a billion years ago.

On the other hand, if a star is very old it is also unlikely to harbor intelligent life. It might seem at first that old stars should offer the best odds for intelligence, because surely the longer the time available for the evolution of intelligence, the more likely it is to appear. But, surprisingly, very old stars turn out to be as poor prospects for intelligent life as very young stars.

The reason is that old stars are deficient in certain important chemical elements, such as carbon, nitrogen, and oxygen, that are necessary for the evolution of life. Since a star and its family of planets are formed from the same cloud of matter, if the star is deficient in carbon and other critical elements, the planets circling around it must also be deficient in these elements. And without the critical elements, life as we know it cannot arise on those planets.

But younger stars and their families of planets—the

175

age of the sun and the earth, for example—do possess the elements needed for life. Our presence in this solar system is proof of that. Where did these younger stars obtain their supply of the life-giving elements?

The answer is that these elements were made in stars. When the Universe first came into existence in the flash of light and heat astronomers call the Big Bang, it consisted almost entirely of hydrogen and helium—the primordial elements of the Cosmos. In that very early period, there were still no stars in the Universe. Later, when the first stars appeared, they were formed from the materials at hand, and so they, too, were made almost entirely of hydrogen and helium.

But with the appearance of these first stars in the Universe, a very important chain of events commenced. Gradually, nuclear reactions within the bodies of the newly formed stars began to transform the light substances, hydrogen and helium, into heavier substances. Among these heavier substances were the life-giving elements, carbon, oxygen, iron, and others.

When the most massive of the early stars came to the ends of their lives, they exploded, spraying into space the carbon and other elements that had been manufactured within their bodies during their lifetimes. Blending with the primordial gases of hydrogen and helium, the freshly distributed elements formed an enriched mixture, which now included, for the first time, the critical elements that make life possible.

In the first few billion years of the Universe's existence, the critical elements were present in very small amounts—for few stars had yet lived, died, and exploded— and the mixture was lean. Then, as more stars exploded and sprayed their contents into space, the mixture became richer. With the passage of time, the abundance of the life-giving heavier elements in the Universe increased further. By the time the sun and its family of planets formed, four and a half billion years ago, carbon, oxygen, and other important substances amounted to two percent of the hydrogen and helium in the Universe.

Two percent does not seem like very much, but it was enough to provide the elements needed for the formation of the earth and, a little later, for the evolution of life on the earth.

Other stars of roughly the same age as the sun—that is, born four or five billion years ago—also contain approximately two percent of the critical elements, because these contemporaries of the sun were formed from cosmic matter of the same richness. Earthlike planets could have formed around those stars, and life may have evolved on these planets.

And stars that were born, say, only one or two billion years ago—that is, stars younger than the sun—must have more than two percent of the important elements, for the cosmic mixture keeps getting richer all the time, as more and more stars explode at the ends of their lives and spray their innards into space. These younger stars, and the planets that circle around them, must also have an ample supply of the ingredients of life.

But stars that formed long before the sun—stars considerably older than four and a half billion years—are less favorable for the evolution of life. When these very old stars came into existence, the Universe was still quite young. At that time, not many stars had lived and died and contributed the materials of their bodies to the Universe. Consequently, in that early period the entire Universe was deficient in the elements needed for life, and any star or planet that formed in that time was also deficient in these elements.

Of course, a star only moderately older than the sun—

say, a billion years or so older—will have only a moderately smaller proportion of the critical elements needed for life. Such a star and its planets would be nearly as suitable for the evolution of life and intelligence as our solar system.

But if the star is very much older than the sun—for example, if it is 13 or 14 billion years old, or almost as old as the Universe—it must be quite deficient in the critical elements, and cannot harbor life.

Therefore, as we examine the stars around us for their suitability as abodes of intelligent life, we may keep a star on our list if it is roughly the same age as the sun or a little older, but a very old star—one that dates back to the early years of the Universe—must be discarded.

And, of course, if the star is very much younger than the sun, we must also discard it from our list, because even if life has evolved on a planet circling that star, this life has not yet had time to evolve to the level of intelligence.

Finally, a star with promising prospects for intelligent life must also be of the right size and mass. If a star is substantially larger and more massive than the sun, it will burn out quickly and come to the end of its life before life can start; or, if it lasts long enough for life to start, it may burn out before intelligence can appear.

The lifetime of a star depends very sensitively on its mass. The sun will live for ten billion years, but stars twice as massive as the sun live for only one billion years—long enough for life to start, but probably not long enough for that life to reach the level of intelligence. And a star ten times as massive as the sun lives for only ten million years—probably too short a time for the evolution of any kind of life.*

*It might be expected that since large and massive stars have more fuel to feed into their nuclear fires, they must last for a longer

Stars less massive than the sun may also be unsuitable for life and intelligence, although here the scientific considerations are less clear-cut. As potential abodes of life, these modest-sized stars have the drawback that their surfaces are cooler than the surface of the sun. Consequently, they do not emit much visible light, that is, radiation with wavelengths the eye can perceive, nearly all their energy being emitted instead in the form of invisible infrared rays. But if a star does not emit visible light, it is hard to see how plants—one of the first rungs on the ladder of life—can evolve. Photosynthesis—the absorption of light by a plant to provide the energy for its growth—is only effective when the plant is exposed to visible light. If a plant is exposed to infrared rays, they do not help its growth at all.

Small stars also emit very little ultraviolet light—a type of radiation that is believed to play an important role in the chemical reactions leading from nonlife to life on a young planet.

On the earth, the evolution of plants led to animals, and the evolution of animals led to intelligent life. Perhaps it is possible for a kind of life even simpler than plants to appear on a planet circling a cool, dim star; but according to our understanding of the history of life on the earth, this life is not likely to evolve to a higher level.

So now we know the stars from whose directions we

time than small stars. But the contrary is true. The reason is that the nuclear reactions which burn up the star's fuel are exceedingly sensitive to temperature. Although a big star has more fuel to burn, the temperature at its center is also higher, because of its greater mass. Consequently, the nuclear reactions inside the star proceed at a rapid rate; when a fire is hotter, the fuel burns faster. As a result, although the large star has more fuel, it burns up its fuel and comes to the end of its life in a shorter time.

are most likely to hear the first extraterrestrial voices in the near future. These promising stars should, of course, have planets; they should be similar to the sun in size and the kind of light they emit; the promising stars should also be about as old as the sun, within a billion years or so; and they should be close neighbors of the sun; in fact, they should be within roughly a dozen light-years from the sun, if we want them to know we are here—and awaiting their call.

13

Stars of Promise

Twenty stars lie within 12 light-years of the sun. Which show the greatest promise as abodes of intelligent life? Let us start with the sun's closest neighbor, Alpha Centauri. This star is at a distance of a little more than four light-years, or 25 trillion miles. The residents of planets orbiting Alpha Centauri, if they exist, have already received the television and radar signals that left the earth in the 1960s. By now they have had more than two decades to listen to these signals and reflect on their significance.

Alpha Centauri happens to be a triple star—three stars formed out of one cloud of matter. The three stars circle around one another in a tight waltz, held in the grip of their mutual gravitational attraction. In general, triple stars (and double stars also) are poor bets for harboring life because they cannot keep their planets in orbit. If a planet forms around one of the stars in a triple star, for example, the gravitational attraction of the other stars tends to pull the planet out of its orbit and send it off into space.

But Alpha Centauri may be an exception because its three stars are separated by relatively great distances. A planet like the earth, circling in a close orbit around one of the stars in Alpha Centauri, probably would not be dislodged by the pull of the other two because they are so far away.

Assuming, then, that the stars in Alpha Centauri are able to keep their planets, which of the three is most likely to have planets that are inhabited?

One of the three stars in Alpha Centauri is considerably smaller and cooler than the sun. It emits very little visible light, even less ultraviolet light, and is not a good prospect for life as we know it. The other two are more promising. They appear to be about the same age as the sun, which means they are old enough so that life could have appeared and could have evolved to an advanced level. Furthermore, one of the two, called Alpha Centauri A, is almost identical to the sun in its size and in the kind of light it emits. This star is a good candidate for the evolution of intelligent life—provided it has earthlike planets.

The next star beyond Alpha Centauri is Barnard's Star, six light-years, or 36 trillion miles, away. Barnard's Star appears to have a family of planets. However, this star is very old—nearly as old as the Universe—and, therefore, is deficient in most of the essential chemical ingredients of life. It is a very poor prospect.

Beyond Barnard's Star lie a number of others, all unpromising for life and intelligence because they are either too small and cool to emit the kind of light needed for life as we know it, or too old to possess the chemical ingredients of life, or too young for intelligent life to have yet appeared. Not until we have traveled nearly 11 light-years from the sun do we come to another star that may harbor life and intelligent beings.

This star is called Epsilon Eridani. It has roughly half the sun's mass and emits somewhat less visible light and ultraviolet light than the sun—but probably enough to allow life to get started. Epsilon Eridani seems to be somewhat younger than the sun—old enough perhaps to have seen the evolution of simple forms of life, but not old enough for the evolution of intelligence. However, measurements of the ages of stars of this kind are accurate only to within a few billion years. When allowance is made for that degree of uncertainty in the age of Epsilon Eridani, this star remains a possible site for the presence of intelligent life.

Nine stars lie beyond Epsilon Eridani but still within 12 light-years of the sun. All but one are too young, too old, too small, or too large to be the abodes of life and intelligence. The exception is Tau Ceti.

Tau Ceti, which is just under 12 light-years from us, satisfies all the basic requirements for the evolution of intelligent life: It is a single star like the sun and—unlike Alpha Centauri—would have no difficulty in holding on to its planets. It is also the same age as the sun, is the same size—and there are signs that it possesses a family of planets. Among all the sun's near neighbors, Tau Ceti is the star most likely to harbor a technologically advanced civilization.

Now the survey of nearby stars is completed. Twenty stars lie within 12 light-years of the sun; three stars of the 20 are places where intelligent life might evolve and flourish. Clearly the sun and its family of planets are also a

183

promising site for life. That means four stars out of 21 in this corner of the galaxy, or roughly one star in five, offer conditions favorable to the evolution of life and intelligence.

Since the galaxy has some 200 billion stars, that means, in turn, that 40 billion stars in our galaxy may harbor intelligent beings. Still another 40 billion or so promising stars may harbor intelligent life in the Andromeda Galaxy, our nearest galactic neighbor. And many other galaxies, each with billions of promising stars, lie beyond Andromeda.

These numbers are staggering. Are all those promising stars actually inhabited? Or only a few? Or none, except the sun and its family?

Some astronomers argue that science already knows the answer to those questions. They reason that the earth is an ordinary kind of planet, containing ordinary materials found in many solar systems. If life emerged on this garden variety of planet, these astronomers ask, why would it not emerge on similar planets elsewhere? Why would the earth alone, among all these planets of the same kind, be chosen by nature or the Deity as the only planet on whose soil the seed of life can take root?

Scientists interested in extraterrestrial life call this reasoning the Principle of Mediocrity. The logic of the Principle of Mediocrity seems impeccable: What better evidence can be found that life and intelligence are common in the Universe than the fact that they have evolved on such a common sort of planet as the earth?

My feeling is that the scientists who believe life is abundant in the Universe are probably right; I also believe in the Principle of Mediocrity. But there is no firm scientific evidence for this view at the present time. My belief in the prevalence of intelligent life in the Universe is more an article of faith—an expression of a personal point of view regarding mankind's place in the Cosmos.

Other scientists, including some of great distinction, are uncomfortable with the Principle of Mediocrity. They point out that while science may be correct in saying that life can arise out of nonliving matter, this may be an event that happens very rarely. After all, the evolution of intelligent beings out of simple molecules is a remarkable development, even if it can be explained by a chain of chemical reactions. It is possible that this extraordinary occurrence requires the concatenation of so many unlikely events as to make the probability of the entire chain essentially zero.

While we know the fraction of stars—roughly one in five—that holds the promise of life, we do not know how many of those promising stars actually are inhabited. Until evidence is obtained on this question, science must accept the possibility that while billions of stars *may* bear life, only one star—the sun—actually does so. The evolution of intelligent life may have happened only once in the galaxy, and perhaps, only once in the Universe. The existence of the human being may be scientifically explainable, and yet, an event so improbable as to be a miracle.

But if a message is received from one of the stars in the sun's close neighborhood, we will know at last that the evolution of intelligent life is not a miracle. We will know that man is not alone; we will know that intelligent beings are probably to be found on many stars in the Universe.

Why is that? How can a conclusion about life on many stars be drawn from the discovery of extraterrestrial life on one star? Consider the three nearby stars—Alpha Centauri, Epsilon Eridani, and Tau Ceti. All three show some promise of harboring intelligent life. Suppose a message was received from one of the three, indicating that it is, in fact, inhabited by an intelligent society. And the sun, of course, is already known to bear intelligent life. That would mean that out of four stars—the sun and three neighbors—which held the promise of life, two have been found actually to harbor intelligent life. Therefore, the probability of finding intelligent life on a promising star is roughly two in four, or 50 percent.

Of course, four stars are too few to give an accurate idea of this important probability. Perhaps if thousands of stars were examined for signs of life, a more accurate value for the probability would turn out to be somewhat higher or lower than 50 percent. But if intelligent life is discovered on one of the sun's closest neighbors, we can be reasonably certain that the probability of intelligent life evolving on a promising star is not as low as, say, one in a million or one in a trillion. For if the probability were as low as this, stars bearing intelligent life would be very sparsely distributed in the sky. It would be most unlikely, in that case, that another star with intelligent life would be found among the sun's nearest neighbors.

So, high stakes ride on the search for messages from nearby stars. If a voice is heard to speak to us by radio from a neighboring star, we will know that the evolution of intelligence is not a relatively rare occurrence in nature, and the Universe probably teems with intelligent life of all shapes, sizes, and levels of development. Over the past few decades astronomers have tuned their radio antennae increasingly to the stars in the hope of catching a whisper, a cry, a lengthy message, that bares the secrets of the Universe to those who can hear. Soon, driven by this hope, they will begin to listen even more carefully.

14

Reaching Out to Cosmic Life

The search for signs of extraterrestrial life began in 1960 in the mountains of West Virginia when a young astronomer named Frank Drake turned the 85-foot radio telescope of the National Radio Astronomy Observatory in the direction of the star Tau Ceti, one of three nearby stars that resemble the sun.

Dr. Drake listened through the night and into the following day, but heard no unusual signals. When Tau Ceti dropped below the horizon, he turned the radio telescope to Epsilon Eridani, only ten light-years away, and also a promising site for intelligent life.

A loudspeaker had been hooked up to the radio receiver so that the astronomers could hear the sounds of the Universe directly. They turned on the equipment. Almost immediately a series of sharp bursts came out of the speaker—*dit dit dit dit dit*—eight times a second. Then the signal stopped. Dr. Drake reports that he and his colleagues felt "great excitement." Was it really this easy to detect extraterrestrial life? They continued to listen carefully to Epsilon Eridani, but the signal from the star was not heard again.

The signal finally reappeared two weeks later. By that time, the astronomers had set up a second radio antenna that was far less sensitive than the 85-foot radio telescope, but could pick up radio signals coming from all directions, and not just the direction of the star Epsilon Eridani. They reasoned that if the burst of sharp sounds was picked up by this antenna also, that would mean it was probably coming from a nearby object, such as a passing truck, or an airplane. The signal did appear on the second antenna, growing stronger and then fading, as if it came from a plane flying overhead. The astronomers' excitement diminished. They had not discovered extraterrestrial life after all.

Other searches for messages from space have been made since then, all with negative results. The task is not easy, because an extraterrestrial civilization might broadcast on any one of many billions of different radio frequencies or channels. Which frequency should the astronomers tune to? They must try every one, because they do not know the frequencies the extraterrestrial beings prefer for their broadcasts.

Suppose the astronomers point their radio telescope in the direction of a promising star and listen to one frequency after another, until they have covered all the billions of possibilities. If they spent only ten seconds listening on each frequency, their search for signals from that one star would take 3,000 years.

Some physicists say the search need not be so time-consuming. They point to a unique frequency in the Universe—a special channel that would be prominent in the thinking of any technically advanced beings as they contemplate their possible choices of frequencies for broadcasting. These physicists point out that space is filled with hydrogen atoms, and every hydrogen atom acts as a tiny radio transmitter, emitting a feeble but very pure signal at a frequency of 1.4 billion cycles per second. Because hydrogen is the most abundant element in the Universe, the signal broadcast by a single atom may be very weak, but the enormous number of hydrogen atoms makes their combined signal very strong, and easily detected by anyone in the galaxy.

Every civilization advanced enough to have evolved astronomers would know that hydrogen atoms broadcast at the special frequency of 1.4 billion cycles per second.* Extraterrestrial physicists in that civilization, pondering their choice of the best frequency on which to broadcast to neighboring stars, might have the same thought our physicists have had: Why not cluster all broadcasts near a frequency of 1.4 billion cycles per second, on the assumption that other scientists in the galaxy would also have the same idea?

The terrestrial physicists concluded that the frequency of the hydrogen atom, or some frequency close to it, is likely to be the channel used by all scientists in the galaxy for communications from one star to another.

This suggestion was helpful to astronomers listening for extraterrestrial messages, because it cuts the number of frequencies, or "stations," to be monitored by a factor of

*Radio and television signals are ripples or wavelets of electric and magnetic force in space that spread out from a radio or TV transmitter like ripples spreading out across the surface of a pond after a pebble has been dropped in. Physicists call these ripples of force electromagnetic waves. The frequency of the signal is the number of ripples or waves that go past a given point each second.

ten, and reduces the listening time for each star from 3,000 years to 300 years.

But monitoring just the three promising stars in the sun's neighborhood would take 900 years—still a long time for one scientific experiment. Even this lengthy period of listening would not be enough to settle the question of intelligent life in the entire Universe, because it may happen that life is common in the Universe as a whole, but, by a fluke, the three particular stars we listen to first may be without life; or one of the three may be inhabited, but its stations may be off the air when we turn our telescopes toward it.

If astronomers listen to three stars near the sun for a number of years and hear nothing from them, they will surely want to go on to listen to other promising stars lying a little farther out. Unfortunately, the number of promising stars increases very rapidly as the search widens. If astronomers go out to 80 light-years—still not very far in the scale of galactic distances—they must add nearly 800 promising stars. The task seems insuperable.

Nonetheless, some years ago NASA asked Congress for funds to begin the search for signs of intelligent life in other solar systems. At first Senator William Proxmire, of Wisconsin, objected because, he said, it was hard enough to find intelligent life in Washington. But later the senator relented, and the space agency began to make preparations for listening to extraterrestrial voices. Since then, scientists at NASA and in several universities have been hard at work building new equipment to cut the listening project down to a manageable size.

Astronomers and other scientists interested in this project call it SETI—the Search for Extraterrestrial Intelligence. They are optimistic about the success of SETI— although some have likened it to looking for a needle in a cosmic haystack—because improvements in electronic equipment have changed the picture since the pioneering search by Dr. Drake in 1960. The radio receiver used by astronomers in that 1960 search could only be tuned to one frequency at a time, and progress was very slow. Today, advances in computers and electronics make it possible to listen to millions of frequencies simultaneously. And the radio receivers designed by NASA specially for SETI can be tuned to *eight million* frequencies at one time.

That 8-million-fold improvement means that the search carried out in 1960, which lasted for two months, can be repeated in a small fraction of a second.

SETI scientists will use the NASA equipment to tune in on nearly a thousand stars. They will listen to each one at 800 million different frequencies. Because of the remarkable new radio receivers, this search will take only a day for one star, and three years of round-the-clock listening for all thousand stars on the list. Since the radio telescopes are also used for other projects in astronomy, the search will take longer than three years. It will begin around 1990, take most of the next decade, and be completed by the year 2000.

This will be the "targeted" search—targeted on stars with special promise for harboring life. But truly advanced beings, a billion years or more beyond humans, may not live on stars or planets. Some astronomers believe that looking for advanced extraterrestrial life on a planet is like looking inside an empty eggshell for a bird. Perhaps intelligent beings hatch on earthlike planets, but when they reach maturity, they fly away.

With these larger possibilities in mind, SETI scientists plan a second listening project called the "all-sky" search. Sweeping their radio telescopes across the heavens, they will listen for messages coming from any direction, even if

191

RADIO TELESCOPES. The radio "telescopes" that listen now and then for messages from extraterrestrial beings are not telescopes in the ordinary sense. They are devices that detect radio waves and measure their intensity and the direction from which the waves are coming.

Stars and galaxies give off many kinds of radiation, including X rays, gamma rays, ultraviolet rays, visible light, infrared radiation, and radio waves. Of these various kinds of radiation, only two— visible light and radio waves—pass freely through the earth's atmosphere. Astronomers describe this situation by saying that the atmosphere has two "windows."

Until 1931, astronomical observations were confined to the visible window, but in that year an electrical engineer named Karl Jansky stumbled across a phenomenon that led to the discovery of the radio window and the birth of radio astronomy. Jansky was not thinking about astronomy at the time, but trying to solve a practical problem in telephone communications. Telephone conversations between London and New York were frequently interrupted by static, and Jansky had the assignment of finding out where it came from. After he eliminated obvious causes of static, such as thunderstorms, one type of noise remained—a hissing sound that did not seem to be connected with any earthly phenomenon.

Probing into the nature of the mysterious hiss, Jansky found that it came from a fixed direction in space, which later turned out to be the center of our Galaxy. Jansky had stumbled on radio waves from space.

Jansky's discovery was made with a few wires stretched on wooden frames. Today most radio astronomers use large, curved reflectors that collect and focus radio waves, just as ordinary telescopes collect and focus light waves.

The largest radio telescope in the world—1,000 feet in diameter— is located in Puerto Rico, in a natural bowl in the hills near Arecibo, on the western end of the island. The photograph (opposite) shows this great "dish," which is immovable but looks in different directions by changing the location of the radio receiver at its focus. The direction in which the telescope points also varies during the course of the day, as the earth rotates. The combined variation is sufficiently great for the instrument to be "pointed" at all the planets in the solar system as well as many galaxies.

no promising stars can be seen in that part of the sky. The scientists plan to divide the sky into one million locations, pausing methodically to listen at each location for about two minutes before moving on to the next one.

They will also cover all radio frequencies that can penetrate through the atmosphere to the earth, and not just those near the frequency of the hydrogen atom. That will allow for the possibility that extraterrestrial physicists are less intrigued by the special frequency of the hydrogen atom than physicists on the earth have been.

Suppose the astronomers receive a signal; how will they tell it is a message from civilization on another star, and not a burst of radio static from the galaxy? The answer has to do with frequencies again. The "signals" produced by natural phenomena are nearly always spread over a broad range of frequencies, but messages sent by intelligent beings will almost certainly be concentrated in a very narrow frequency band. (If the broadcasters spread their signal over a broad band of frequencies, the signal will be too weak at any one frequency to stand out against the natural background of galactic static.) When the astronomers tune their radio receivers to the frequency on which the extraterrestrial scientists are broadcasting, the signal will pop out sharply.

One can imagine the radio astronomer standing in front of the receiver, slowly turning the dial, tuning in to one frequency after another, and hearing only static—until, at one particular point on the dial, a sharp, clear signal is heard. That is just what the NASA and university scientists will be doing, except that a computer will turn the dial, tuning in on eight million frequencies at a time.

•

We have begun to listen. Soon we will listen harder. Eventually, the message will come—unless mankind is alone. As we contemplate this possibility, we must remember that when the eagerly awaited message arrives, it is likely to be from a nonhuman form of intelligent life—creatures as far removed from humans in the evolutionary scale as humans are removed from the crawling creatures of the sea.

Is there value in making contact with a form of life so far above us? Is it wise to do so? For a hint of the rewards from such an encounter, consider a dialogue with creatures on a planet a mere million years older than the earth.

A million years is a long time in the evolution of intelligence. Great scientists like Einstein seem to appear every thousand years; great inventors of new technologies, perhaps every few decades. In the next million years, a thousand Einsteins may live on the earth, and hundreds of thousands of great inventors. Suppose an older society, which has long since traveled the path we are just starting on, made all these theories and inventions available to us. We might learn the secret of immortality, of virtually unlimited energy, of cures for the most dreaded diseases. The wisdom of an older and far superior civilization could become ours.

The thought of such riches staggers the imagination. But there are risks. In the contact between scientifically advanced civilizations and a primitive society—and such is the description we must apply to humans as they prepare to enter the galactic community—it is the usual lot of the less developed peoples to be destroyed by the encounter. Regardless of the intent of the technically advanced civilization, powerful forces at their command tear apart the fabric of the primitive society. Such was the fate of the

American Indian, the Australian aborigines, Tasmanians, and Polynesians.

These have been the fateful consequences of exposure to a more advanced technology when some thousands of years of cultural evolution separate the two societies. What may be expected from a meeting between civilizations separated by many millions of years of evolution? Can mankind survive the shock of the encounter?

It is too late for these thoughts. The shell of television broadcasts, moving outward from the earth, has swept past many stars like the sun. They know we are here. There is no turning back.

Afterword

The Evolving
Perception of the
Cosmos

The earth seems vast and immobile; throughout two million years in the prehistory of man it provided the stage for all human experience, with the heavens seemingly no more than a backdrop of moving lights. Astronomy and the exploration of space have led us to the contrary realization that the Universe is vast, and the world of men is small.

•

To anyone who follows the motion of the sun day after day, and the motions of the moon and the stars night after night, it is obvious that the earth is the center of the Universe and that the heavenly bodies revolve about it daily, paying homage to the abode of man. Every day the sun moves across the vault of the heavens; every night the moon and stars travel in their stately procession across the sky.

In ancient times, men marveled at this nightly movement of the heavenly bodies and wondered what its cause could be. As they followed the stars from night to night, they saw further that their shapes were unchanging; the stars of the Big Dipper moved across the sky as a unit then, and still do today. Reflecting on these facts, the early astronomers decided that the stars must be attached firmly to an enormous sphere surrounding the earth. The sphere revolved around the earth every 24 hours; as it turned, the stars turned with it. In the center of the sphere of stars rested the earth, solid and unmoving, fittingly placed at the hub of the Universe.

A few astronomers in ancient Greece reasoned that it might be the earth, and not the heavens, that turned on its axis every 24 hours. That would create an apparent motion of the sky. The stars could be fixed in space, but a person on the rotating earth would see them moving past him in the opposite direction, just as the landscape seems to move past a person on a merry-go-round. And one Greek astronomer had an even stranger thought; he proposed that the earth might be moving around the sun, in addition to turning on its axis.

Although these ideas are familiar to us today, in their time they were ridiculed or ignored, and after a while they disappeared entirely. It seemed nonsensical to most people in early times to postulate that the massive earth could spin on its axis like a top, or sail through space like a ship. Surely, everything not fastened to the earth would be left behind; an arrow shot straight up into the air would come down many miles away; a stone dropped from a tower would never reach the ground beneath; rocks and trees would be hurled from the spinning earth, like mud flung from the rim of a turning wagon wheel. Since no one ever saw such effects, the earth must be stationary, and the sun, moon, and stars must revolve around it daily. All human experience proved that this is so.

But one fact did not agree with the picture of a fixed earth surrounded by a rotating sphere of stars. Five "stars" did not behave like ordinary stars; instead of maintaining fixed positions relative to other stars, they wandered about in the heavens, at times near one star and at times near another. The Greek astronomers, puzzling over the fact that the five mysterious objects were unlike any other stars, called them the Wanderers, or *planētēs*, in Greek. In English, they became known as the planets.

Today, a planet means a massive ball of rock and iron like the earth or Mars, or a huge sphere of hydrogen like Jupiter; but the Greek astronomers, lacking telescopes, had no idea that the objects they called planets could be massive bodies like the earth. To them, a planet was just a starlike point of light in the sky.

Bemused by the erratic motion of the planets, the early astronomers observed their positions carefully, year after year, and after a time they perceived a pattern in the movements. Each planet, or wandering star, followed a looping path in the night sky, first curving around from east to west, and then back from west to east.

If the planets were attached to a great sphere that revolved in the heavens, they would move across the sky only from east to west, in a fixed procession with the rest of the stars. Clearly, they could not be attached to the celestial sphere. They must be located somewhere else in space. But where? And why did they move back and forth in loops?

Reflecting on these questions, two Greek astronomers named Apollonius and Hipparchus had an ingenious idea. They reasoned that each planet must be attached to the rim of a wheel that rolled across the sky. As the wheel rolled across the sky, the planet would describe a looping path, just as the planets are observed to do.

The idea worked well in principle. However, when the astronomers tried to make their picture of wheels rolling across the heavens agree with the observed motions of the planets, they found that an accurate fit was impossible unless they assumed that wheels were rolling on wheels. That is, a planet moved on the rim of one wheel, which moved around the rim of still another wheel.

The astronomer Ptolemy, who lived in the second century A.D., found that no less than 40 wheels on wheels were needed to describe the movements of the sun, the moon, and the five planets known at that time.

Ptolemy's rolling wheels worked well, but people found his model of the heavens to be very complicated. When Alfonso, king of Castile and Aragon, heard about the Ptolemaic system, he remarked, "If the Lord had consulted with me, I would have recommended something simpler." And John Milton, who had to teach the Ptolemaic system as a schoolmaster in the seventeenth century, wrote in disgust about Ptolemy and his followers,

> How they contrive
> To save appearances, how gird the Sphere
> With Centric and Eccentric scribbled o'er
> Cycle and Epicycle, Orb in Orb . . .

Still, Ptolemy's picture of the Universe was the best the human brain could devise. In spite of its cumbersome machinery, it lasted for nearly 14 centuries.

Finally, around 1500, a Polish churchman named Copernicus made a strange proposal. The earth moves, he said, reverting to the discredited and forgotten theories of the ancient Greek astronomers. The earth moves around the sun every year, and turns on its axis every day. No longer is the earth at the center of the Universe. Now it is the sun, Copernicus wrote, that "sits in the midst of all enthroned," the sun, "rightly called the Lamp, the Mind, the Ruler of the Universe, gathering his children the Planets which circle around him."

Why did Copernicus resurrect ideas that had died with ancient Greece—ideas that had been dead for nearly 2,000 years? He described his reasons later, in the introduction to his book on the new theory of the Universe. It seemed more reasonable, he wrote, to assume the earth turned on its axis every 24 hours than to believe the entire Universe traveled with incredible speed in the opposite direction every day.

But most astronomers regarded the theory of Copernicus as a failure. All the old problems remained. If the earth were turning, objects and people would fly off it. If it moved around the sun, things would be left behind in space. "Fool," Martin Luther called Copernicus, for trying to turn astronomy on its ear, and contradicting the Bible to boot; for as Luther angrily pointed out, "The Holy Scriptures tell us that Joshua commanded the sun to stand still, and not the earth."

Yet, gradually the Copernican theory took root in men's minds. There was an air of freshness about it; it opened the door on new ideas that went far beyond the science of astronomy. Men saw the implications in the sun-centered Universe. Consider the following facts, they said: The five planets revolve around the sun; the earth revolves around the sun; since the planets behave in the same way as the earth, they must be similar objects.

This was a surprising conclusion. Prior to that time, astronomers had believed that the planets were hard, polished spheres, made of a jewellike substance, perfect and unchanging, while the earth was made of ordinary substances like mud, rock, and water. If the earth and the planets are similar objects, the followers of Copernicus reasoned, the planets may also be made of mud, rock, and water. That unsettling thought led to another: If the planets are made of the same material as the earth, perhaps they bear life; perhaps there are people on them.

Those were radical ideas. They presaged the latest developments in twentieth century science, which unite life on the earth to life in the Cosmos. The astronomy of Copernicus was the first step in the Copernican revolution; Copernicus removed the earth from the center of the Universe, and put the sun in its place. Others took the second step; they removed the sun from the center and put nothing in its place. There is no center, they said; the Universe is infinite, and contains an infinite number of stars. Each star is a sun like ours; each has a family of planets.

These ideas led finally to the modern picture of a Universe populated by innumerable suns, innumerable earths, and, perhaps, innumerable forms of life. That thought expresses the essence of the Copernican revolution: Mankind is supreme among the forms of life on this planet, but in the cosmic order his position is humble. No revelation more striking has ever come from the scientific mind.

Acknowledgments

I am indebted to many friends and colleagues who have given generously of their time in conversations and correspondence relating to the technical aspects of space exploration and the properties of the terrestrial planets. Barney B. Roberts of the NASA Johnson Space Center provided a valuable overview of the planning for manned flights to Mars and the limiting technical factors in long-duration space flight. Dr. Arnauld E. Nicogossian of NASA Headquarters contributed fascinating information on the medical and psychological hazards in year-long flights across the solar system. Professor John Lewis of the Lunar and Planetary Laboratory of the University of Arizona shared with me his enormous fund of theoretical and observational knowledge regarding the properties of the major and minor bodies of the solar system and the origin and evolution of planetary atmospheres.

James Oberg was, as always, an invaluable source of information on Soviet plans and accomplishments in space flight, and particularly on the Soviet Phobos and Mars missions in relation to U.S. plans for Mars exploration. I am also grateful to Duke Reiber for his help in pulling together a large volume of information derived from the NASA Mars Conference. Professor Ben Finney of the University of Hawaii contributed insights into the history of human exploration which enriched my understanding of the prospects for flights outside our solar system.

It has been a pleasure to work with Henry Ferris, my editor at Bantam Doubleday Dell. His highly intelligent criticisms and suggestions improved the organization of the entire work, as well as the clarity of individual passages. Doris Cook continued to be my valued partner in this work, as she has been in all my writing during the past ten years. Her excellent sense of language has contributed much to the book's explanations of scientific concepts. I am indebted to my good friend Lisl Cade for stimulating discussions of all the major themes in the book. My mother, Marie Jastrow, made many suggestions which further improved the accessibility of technical passages for the general reader. The book owes a great deal to her clarity of thought.

Index

Page numbers in italics refer to illustrations

207

Photo Credits

Chapter 5
  75 Copyright © 1986, Royal Observatory, Edinburgh, courtesy of Lance Miller
  80 The National Radio Astronomy Observatory, courtesy of Richard A. Perley, John W. Dreher and John J. Cohan
  81 Hale Observatories

Chapter 6
  99 Courtesy of Barrett Gallagher
  100–101 Jet Propulsion Laboratory, California Institute of Technology
  104–105 U.S. Air Force
  106 Lunar and Planetary Laboratory, Department of Planetary Sciences, University of Arizona, courtesy of Uwe Fink
  107 Bell Aerospace Corporation © 1986, Max Planck Institut für Aeronomie, courtesy of Harold Reitsema

Chapter 7
  114 Jet Propulsion Laboratory and University of Arizona, courtesy of Bradford Smith and Richard Terrile
  115 Hale Observatories

Chapter 8
  121 United Press International
  123 Jet Propulsion Laboratory
  130–135 Jet Propulsion Laboratory

Chapter 9
  146–147 Jet Propulsion Laboratory

Chapter 10
  151 NASA Historian's Office, NASA Wallops Flight Facility

Chapter 14
  193 Arecibo Observatory (National Astronomy and Ionosphere Center, Cornell University)

Printed in the United States
by Baker & Taylor Publisher Services